LONG-TERM PROCESS-BASED MORPHOLOGICAL
MODELING OF LARGE TIDAL BASINS

Long-term process-based morphological modeling of large tidal basins

DISSERTATION

Submitted in fulfillment of the requirements of
the Board for Doctorates of Delft University of Technology
and of
the Academic Board of the UNESCO-IHE Institute for Water Education
for the Degree of DOCTOR
to be defended in public,
on Monday, 12 November 2012, at 10:00 o'clock
in Delft, The Netherlands

by

Ali DASTGHEIB

born in Shiraz, Iran
Bachelor of Science in Civil Engineering, Shiraz University, Iran
Master of Science in Civil Engineering, Amir Kabir University, Iran
Master of Science in Hydraulic Engineering, UNESCO-IHE, The Netherlands

This dissertation has been approved by the supervisor:
Prof. dr. ir. J.A. Roelvink

Composition of Doctoral Committee:

Chairman	Rector Magnificus Delft University of Technology
Prof.dr. ir. A. Szöllösi-Nagy	Vice-Chairman, Rector UNESCO-IHE
Prof. dr. ir. J.A. Roelvink,	UNESCO-IHE/Delft University of Technology, supervisor
Prof. dr. Gary Parker	University of Illinois at Urbana, USA
Prof. dr. ir. Z.B. Wang,	Delft University of Technology
Prof. dr. ir. A.E. Mynett,	UNESCO-IHE/Delft University of Technology
Dr. ir. A.B. Fortunato	LNEC, National Civil Engineering Laboratory, Portugal
Dr. A.J.F. van der Spek	Deltares
Prof. dr. ir. M. Stive,	Delft University of Technology, reserve

CRC Press/Balkema is an imprint of the Taylor & Francis Group, an informa business

© 2012, Ali Dastgheib

Cover photo: Old map of Belgii Foederati Jansoni (1658) together with the results of numerical simulation carried out in the study presented in this book

Published by:
CRC Press/Balkema
PO Box 447, 2300 AK Leiden, the Netherlands
e-mail: Pub.NL@taylorandfrancis.com
www.crcpress.com - www.taylorandfrancis.co.uk - www.ba.balkema.nl

ISBN 978-1-138-00022-3 (Taylor & Francis Group)

Daar is het water, daar is de haven,
Waar j' altijd horen kon: Wij gaan aan boord.
De voerman laat er nou paarden draven
En aan de horizon ligt Emmeloord

Eens ging de zee hier tekeer,
Maar die tijd komt niet weer,
Zuiderzee heet nou IJsselmeer.

- From " Zuiderzeeballade"

Summary

The morphology of tidal basins includes a wide range of features developing along different spatial and temporal scales. Examples are shoals, channels, banks, dunes and ripples. Coastal engineers use their engineering tools to answer questions on the processes governing the short term (< decades) development of these morphological features. Geologists apply their conceptual models and reconstruction methods to answer questions related to a much longer time scale (> centuries). This two-sided approach has left us with limited understanding of processes occurring on intermediate scales (> decades and < centuries), whereas the morphodynamics of these intermediate scales are of special concern to sustainable coastal zone management.

This study is part of a collective effort to bridge the aforementioned gap by extending the use of coastal engineering tools (process-based models) to geological time scales to provide more understanding of the physical processes governing the long-term morphodynamic behavior of tidal basins. A fundamental question addressed is whether or not process-based models can reproduce trustworthy long-term developments. To answer this question the Dutch Waddenzee is chosen as a reference case.

This study suggests that the question has a positive answer. By comparing model results with measured developments in the Waddenzee, this study shows that a process-based model can reproduce channel-shoal patterns and their long-term development qualitatively well. Modeled parameters such as area, volume and height of the inter-tidal flats obey the data-based equilibrium equations. This study also demonstrates the models' ability to qualitatively assess the impact of large scale human intervention in a tidal basin. For example, the model is able to reproduce the change in tidal transport regime and the ensuing morphodynamic changes due to an extreme impact such as the closure of the Zuiderzee.

Although the highly schematized simulations produced qualitatively good results, they also revealed the need for a better process description. As the first step to improve model performance a methodology was developed to account for sediment composition and distribution in the bed. In the next step different methodologies to schematize wave action for long-term morphological simulations were investigated. investigated the wave climate. Model results show that the chronology of wave conditions and the wave schematization approach have a limited effect. The outcome of long-term (decadal) morphodynamic simulations with different wave and tidal conditions are in good agreement with conceptual models. For the reference case, model results revealed that the morphological impact of wind waves is not only important outside the inlet and at the ebb-tidal delta, but also within the tidal basin. A final conclusion is that adding methodologies for bed composition and wave schematization to the model of the Waddenzee area improved the hindcasting simulations qualitatively.

Samenvatting [1]

De morfologie van getij bekkens beschrijft een breed scala aan vormen die zich ontwikkelen volgens een eigen schaal in ruimte en tijd. Voorbeelden zijn platen, geulen, zandbanken, duinen en ribbels. Kustwaterbouwkundige ingenieurs gebruiken hun model concepten om vragen te beantwoorden met betrekking tot korte termijn (< decaden) ontwikkeling van deze morfologische vormen. Geologen passen hun eigen conceptuele modellen en reconstructie methoden toe om vragen te beantwoorden op veel langere tijdschalen (> eeuwen). Door deze twee benaderingen blijft de kennis van tussen liggende tijdschalen (> decaden en < eeuwen) onderbelicht, terwijl deze tijdspanne juist van belang is voor het duurzame beheer van een kustzone.

Deze studie is onderdeel van een gezamenlijke inspanning om het eerder genoemde hiaat in kennis op te vullen. Kustwaterbouwkundige modellen wordt opgerekt naar geologische tijdschalen om meer kennis en begrip te krijgen voor de fysische processen die lange termijn morfodynamisch gedrag van getij bekkens bepalen. Een fundamentele vraag is of proces gebaseerde modellen betrouwbare lange termijn ontwikkelingen kunnen reproduceren. De Nederlandse Waddenzee dient als case studie om deze vraag te beantwoorden.

Deze studie suggereert dat deze vraag een positief antwoord heeft. Door een vergelijking van model resultaten met gemeten ontwikkelingen in de Waddenzee, laat deze studie zien dat een proces gebaseerd model de ontwikkeling van het plaat-geul systeem kwalitatief goed kan reproduceren. Gemodelleerde model parameters zoals het oppervlakte, volume en hoogte van het intergetijde gebied volgen beantwoorden aan empirische, evenwichtrelaties. Deze studie toont ook de potentie van proces gebaseerde modellen aan om de impact van grootschalig menselijk ingrijpen in getij bekkens kwalitatief in te schatten. Het model is bijvoorbeeld in staat om het veranderde getij transport en de resulterende morfodynamische ontwikkeling te reproduceren als gevolg van extreem ingrijpen zoals het afsluiten van de Zuiderzee.

Hoewel de geschematiseerde simulaties tot kwalitatief goede resultaten hebben geleid, toonden ze ook de potentie aan om de model resultaten te verbeteren. Als een eerste stap beschrijft deze studie een methodologie om de initiële sediment verdeling en samenstelling in de bodem te bepalen. Een tweede verbetering betreft de ontwikkeling van een methodologie voor golf schematisatie voor lange termijn morfodynamica. Model resultaten laten zien dat de chronologie van golf condities en de golf schematisatie een beperkt effect hebben. De resultaten van lange termijn (~decaden) simulaties met verschillende golf en getij condities komen goed overeen met conceptuele modellen. Met betrekking tot de case studie laten de model resultaten zien dat de morfologische impact van golven niet alleen belangrijk is zeewaarts van het getij bekken en de ebb delta, maar ook in het bekken zelf. Een belangrijke conclusie is dat de voorgestelde methoden voor bodem samenstelling en golf schematisatie de hindcast van de morfodynamische ontwikkeling van de Waddenzee kwalitatief hebben verbeterd.

---

[1] This summary is translated to Dutch by Dr. Mick van der Wegen, Senior lecturer in UNESCO-IHE.

Acknowledgment

I want to start with acknowledging UNESCO-IHE especially Dano Roelvink, Joop de Schutter and Han Ligetingen for trusting me, hiring me as a lecturer and giving me the opportunity to do my PhD research parallel to my other duties.

Dano is famous for solving 80% of the problems in the numerical modeling in 5 minutes, in supervising my work it was indeed the case, and the remaining 20% took me more than 4 years to solve. Dano you always found time in your busy agenda to answer my questions and guide me in the right path. Thank you for all the time, help and flexibility that you offered me, working with you is a pleasure.

No one can wish for an office-mate better than Mick van der Wegen, intelligent, generous, helpful, inspiring and sailing enthusiast. Thank you Mick for the last 5 years.

My other colleagues in CSEPD chair group of UNESCO-IHE : Pushpa Kumara Dissanayake, Rosh Ranasinghe, Han Ligteringen, Poonam Taneja, Frank van der Meulen, Gerard Dam, Fernanda Achete, Guo Leicheng and Johan Reyns, thanks you very much for the nice working environment. Johan thank you for your help in the last couple of weeks of writing this dissertation.

I want to also acknowledge Deltares for providing the necessary funding for this PhD. This study was funded in the frame work of "Kustlijnzorg" project of Deltares. I want to especially thank Ad van der Spek, Zheng Bing Wang, Ankie Bruens , John de Ronde and Edwin Elias, not only you made this study financially possible but also provided the necessary software, essential data, and helpful comments.

I also should thank Huib de Swart, Bert Buurman and Mohammad Adel, for their ideas and discussions which contributed to this study.

Seven years living in Delft cannot pass without making a few good friends :

Fellipe Gonzalez, Julien Chenet, Andrea Silva, Sylvie Kanimba, Maria Fernanda Jarquin, Dragan Tutulic, Paola Reyes, Elena Benedi and Meshkat Dastgheib, Delft was never the same without you guys.

Stefania Balica and Sergio Chelcea, you are always like family to me. Your occasional visits to Delft and me taking refuge in your home were always a refreshing getaway. Thank you for your generosity and friendship.

Shilp Verma, Thank you for the entire interesting, funny, eye opening, never ending discussions which we had during our numerous trips, let's hope for many more to come.

Luigia Brandimarte, Giuliano di Baldassarre, Alida Pham, Lena Heinrich, Roham Bakhtiar and Maria Rusca, you proved to me that the length of a friendship is not as important as its depth. May we continue to be good friends for a long time.

Carmen Guguta, Carlos Lopez, Mariana Popescu, Pooneh Rezaee, Majid Jandaghi, Assiyeh Tabatabai, Ladan Adili, Denys Villa, Jorge Almarez, Samira Fazllollahi, Amin Karbasi, Amir Razmi and Julliette Cortes. Thank you for all the nice moments that we have shared.

Perhaps I should finish by acknowledging my parents, I want them to know everything that I value the most in my life, they have given me. Mahroo and Masih, your patience, never ending encouragement and unconditional love have supported me each and everyday of my life.

Ali Dastgheib

26th of June 2012
Delft

Table of Contents

1. Introduction   1

1.1   Motivation   2

1.2   Different engineering approaches in long-term morphological modeling   3

   1.2.1   Behavior based models   3

   1.2.2   Data-based models   3

   1.2.3   Process based models   4

   1.2.4   Inverse methodology   5

   1.2.5   Formally integrated, long term models   5

1.3   Goal of this study   5

1.4   Relevance   6

1.5   Research questions   6

1.6   Spectrum of the simulations and the structure of the study   7

2.So far so good : *Long-term process-based morphological modeling of the Marsdiep tidal basin*   9

2.1   Introduction   10

2.2   Objectives   10

2.3   Study area   11

2.4   Model description and setup   13

   2.4.1   Model description   13

   2.4.2   Grids   15

   2.4.3   Forcing   15

   2.4.4   Initial bathymetry   16

   2.4.5   Different Runs   19

2.5   Model results and discussion   19

   2.5.1   Model result for Marsdiep basin   19

   2.5.2   The boundary between different basins   26

   2.5.3   Effect of initial bathymetry   28

2.6   Conclusion   28

3. Turning the tide : *Long-term morphodynamic effects of closure dams on tidal basins*   31

3.1   Introduction   32

3.2   Aim of Study   33

3.3   Study area   33

3.4   Model Description   34

3.5   Model Setup   35

   3.5.1   Model Domain   35

   3.5.2   Forcing   36

   3.5.3   Initial Bathymetry   36

   3.5.4   Scenario of simulations   36

3.6   Results and Discussion   37

   3.6.1   Morphological changes   37

   3.6.2   Effect of closure on the tidal wave propagation   39

   3.6.3   Import / Export regime   39

   3.6.4   Intertidal flat characteristics   41

   3.6.5   Texel ebb-tidal delta development   43

   3.6.6   Boundary of the basins (Tidal divides)   44

3.7   Conclusion   45

4. Mixing the soil : *Effect of sediment see bed composition in long-term morphological modeling of tidal basins*
                                                                                       47

4.1   Introduction   48

4.2   Aim of the study and methodology   49

4.3   Area of interest and available data   49

4.4   Model Description   51

   4.4.1   Bed layer model   51

4.5   Long-Term model set up   52

   4.5.1   Selection of sediment classes and bed layer model setup   53

   4.5.2   The effect of sediment classes on long-term morphodynamic simulations   55

4.6   Realistic hindcast with sediment classes   58

   4.6.1   Preparing the initial sediment distribution for the hindcasting model   60

   4.6.2   Hindcasting Model setup   60

   4.6.3   Bench mark simulations   64

4.7   Results and Discussions   64

   4.7.1   Brier Skill Scores (BSS)   64

   4.7.2   Cross Sections   68

   4.7.3   Effect of using sediment mixtures on the impact of the bed slope on the
   adjustment of the sediment transport rate   69

4.8   Conclusions   70

5.Switching channels : The effect of waves and tides on the morphology of tidal inlet systems   71

5.1   Introduction   72

5.2   Aim of the study   74

5.3   Model Description   75

    5.3.1   FLOW module   75

    5.3.2   WAVE module   75

    5.3.3   Coupling the Modules   76

5.4   Model Setup   77

    5.4.1   Grid and Bathymetry   77

    5.4.2   Sediment and sediment transport relation   78

    5.4.3   Forcing of the model   79

5.5   Different simulations   80

5.6   Results and Dissuasions   81

    5.6.1   Morphology of ebb-tidal delta   82

    5.6.2   Sediment bypassing and cyclic behavior of main channel   83

    5.6.3   Hydrodynamic and sediment transport patterns   89

    5.6.4   Effect of wave stirring   91

5.7   Conclusion   91

6. Unleashing the waves : Wave schematization approaches for long-term morphological modeling of tidal basins   93

6.1   Introduction   94

6.2   Area of interest and available data   94

    6.2.1   Wave data   95

    6.2.2   Tides   97

    6.2.3   Bathymetry   98

    6.2.4   Wind and surge   98

6.3   Model Description   99

6.4   Schematization of forces   99

    6.4.1   Tide   99

    6.4.2   Wave   100

    6.4.3   Wind   106

    6.4.4   Surge   107

    6.4.5   Applying schematized wave conditions together with morphological factor   107

6.5   Effect of Wave Schematization   109

    6.5.1   Model Setup   110

6.5.2	Impact of waves	112
6.5.3	Comparison between approaches of wave schematization	115
6.5.4	Effect of using variable MorFac comparing to a single MorFac	121
6.6	Hindcasting	122
6.6.1	Model setup	122
6.7	Results and Discussions	124
6.8	Conclusions	132
7. Conclusions and recommendations		135
7.1	Response to research questions	136
7.2	Recommendations	138
7.2.1	Input schematization and climate change	138
7.2.2	Sensitivity analyses for virtual reality simulations	138
7.2.3	Three dimensional processes	138
7.2.4	Fine and cohesive sediments	139
7.2.5	Evaluation of morphological simulations	139
Appendix		141

CHAPTER 1

Introduction

1.1 Motivation

The morphodynamics of tidal basins include a wide range of features with different spatial and temporal scales. Coastal engineers use their engineering tools to answer questions on the processes governing the short term (< decade) development of these morphological features. Geologists apply their conceptual models and reconstruction methods to answer questions related to a much longer time scales (> centuries). This two-sided approach has left us with limited understanding of processes occurring on intermediate scales (> decades and < centuries), whereas the morphodynamics of these intermediate scales are of special concern to coastal zone management.

Therefore there is a need to improve the understanding of important processes at intermediate scales as well as the ability to predict the behavior of morphological features with such scales. To bridge this gap either the coastal engineers should try to extend their understanding of the morphological processes and adapt their tools to be used beyond what is used now or geologists should downscale their very large scale models. (Figure 1-1)

In this study the first approach of the abovementioned two is adopted, and it is tried to look towards the intermediate scale morphodynamics of tidal basins with engineering glasses, the immediate consequence of this approach is that the intermediate scale morphodynamics appears to be long and large scale morphodynamics from an engineer point of view.

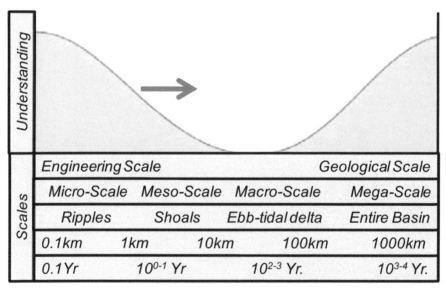

Figure 1 - 1The gap in understanding of the intermediate scale morphodynamics and the approach adopted in this research (Re-drawn after De Groot 1999 courtesy of Ad van der Spek)

1.2   Different engineering approaches in long-term morphological modeling of tidal basins

There are lots of different phenomena affecting the morphological evolution of tidal basins. The effects are varying in a wide range of temporal and spatial scales. Therefore one model type can not provide all the necessary morphological information to fulfill every kind of needs. It is claimed by De Vriend (1996) that "the all purpose model for tidal inlet morphodynamics does not exist and is not likely to emerge in near future". To the knowledge of the author such a model is not developed in recent years either. The models which perform well for small-scale phenomena do not necessarily perform the same on larger scales and vice versa. So, the morphological models also should be classified according to the morphodynamic scale of phenomena of interest. Since long-term modeling is the main approach in this study, different types of long-term models used by coastal engineers are described briefly in this section.

In morphological modeling of the coastal zone, including tidal basins, there are two main approaches described by De Vriend et al. (1993b): 'behavior oriented models' and 'process-based models'. He also introduced a combination of these two approaches called 'formally integrated, long-term models' later (De Vriend, 1996). In recent years Karunarathna et al. (2008, 2009) introduced another approach to combine behavior oriented and process-based models, based on the inverse methodology.

1.2.1   Behavior based models

In behavior based models the physics of underlying processes are neglected and modeling is based on the empirical relation between different coastal phenomena. These kinds of models rely on the available measured data of coastal parameters. Different behavior based models have been developed for long-term morphological modeling including the models specifically aimed at tidal basins. Behavior based models by are also categorized in different classes by De Vriend (1996).

Data-based models

The main assumption in this type of model is that the processes which govern the trend of the evolutions in coastal region remain constant during the time; or in other words in this models it is assumed that the coastal parameters continue their past evolution with the same trend. The simplest form of this model is the extrapolation of a parameter in time using a linear regression. The more sophisticated model of this type using a multi-scale nonlinear system is also developed and has been used in tidal basin systems. Using the relations resulting from regressions the behavior of the parameters can be predicted.

Another type of the data-based model called 'translation' in literature, predicts the behavior of a coastal system by using the data and evolution history of a well monitored similar system. In this type of model it is assumed that both systems respond to the interferences similarly.

The data based models have been used in predicting the behavior of the tidal basins. However in such a complex model there are a large number of inputs and parameters to be taken into account. So it is difficult to find the relation between the mechanisms and different aspects of evolution. Therefore using data-based models in tidal basins requires a good understanding of the physical mechanisms. (De Vriend, 1996).

Empirical models

The empirical models belong to two different categories: the equilibrium state models (relationships) and the transient empirical models.

In the equilibrium state relations, it is assumed that there is a coastal system (e.g. tidal basin) which is already in its (dynamic-) equilibrium which can be chosen as a prototype of other systems with similar conditions. Relations between different parameters of such a system are determined by analyzing the measured data; these relations are assumed to be valid for similar systems. This type of relations has been used in tidal basin morphology extensively. However, in most of the available equilibrium state relations, some of the data, which have been used in the analyses, are not the data from equilibrium conditions.

Transit empirical models describe the evolution of a morphological parameter between a given actual state and its equilibrium state as an exponential decay process (De Vriend, 1996). Transit empirical models are based on the assumption that each element of the system behaves independently. This assumption is not always true for a tidal basin system, in which there are sediment exchange between basin, ebb-tidal delta, and adjacent coastline. In other words this kind of model can be used in a tidal basin system only if sufficient sediment is available inside the modeled element.

Semi-empirical models

In semi-empirical models, the approach is to use all kind of available information, such as measured data in the field, equilibrium state equations, and large scale balance equations based on the available theories. Due to lack of enough detailed empirical information, these models can not be used in detailed scale. In the case of a tidal basin, these models are developed for the scale of basin, ebb-tidal delta, etc.

A large number of models with this approach have been developed for different elements of a tidal basin system, some of them are :   Di Silvio's basin model (Di Silvio, 1989), Van Dongeren's basin model (Van Dongeren and De Vriend, 1994), Karssen's basin model (Karssen, 1994a, 1994b), De Vriend et al's delta model (De Vriend et al, 1989).), Steetzel's model of the entire Waddenzee coast (Steetzel, 1995). More recent models of this type, which are still developing are ESMORPH and ASMITA. (Stive et al. 1998, Stive and Wang 2003)

1.2.2   Process-based models

Process-based models are based on the description of underlying physical processes. This type of models consists of a number of modules which describe different processes such as hydrodynamics (wave and current) and sediment transport. These modules interact dynamically with bathymetry and lead to the morphological changes.

These models need a careful selection of the processes to be modeled. Each of the relevant processes should be modeled adequately, not only in the sense of process description, but also, the combination of the modules, which forms the model as a whole (De Vriend, 1996). On the other hand the input to the model should be schematized. In this regard another distinction is made on the process-based models. Some models simulate the long-term effects, by modeling the full description of the of small scale processes, but with the schematized inputs. Using the morphological tide to represent the full neap spring tidal cycle is an example of such approach. This type of approaches is referred as 'Input reduction' in literature. Some other models only use the modules to describe the most important physical processes. This approach is called 'Model reduction'. However in most cases both input reduction and model reduction concepts are applied.

Another issue in process-based long-term morphological modeling is how to couple the basic modules of different processes in a model, to reach the result with desired accuracy in a reasonable computational time. Roelvink (2006) summarized different techniques which are developed in this regard.

As Roelvink and Reniers (2012) indicated, the other advantage of the process-based models is that not only these models try to reproduce and predict the reality "Virtual reality", but also they can be used as a numerical lab to examine the effect of different processes and find the physical processes which play the main role in different morphological behavior and answer more fundamental equations "Realistic analogue".

In recent years, process-based models have been used in different studies to simulate the morphology of tidal basins and estuaries for different scales, such as Wang et al. (1995), Marciano et al. (2005), Van Leeuwen et al. (2003), Van der Wegen and Roelvink (2008), Van der Wegen et al. (2008, 2010, 2011), Dissanayake et al. (2009), Tung et al. (2008, 2009, 2011)

1.2.3   Inverse methodology

This technique is mainly used for predicting long-term variations in the morphology of estuaries. In this method a morphological evolution equation is used which isolates diffusive and non-diffusive processes in estuaries. The contribution from non-diffusive processes to the morphological changes of the estuary is incorporated in the governing equation by a source function. The source function is derived by solving an inverse problem using historic data and Empirical Orthogonal Function (EOF) analysis is used to analyze the spatial and temporal variation in the source function (Karunarathna et al., 2008). The disadvantages of this method are mainly the need for a long and large data set. The governing equation of the model is case limited and can not be adopted for other cases.

1.2.4   Formally integrated, long-term models

Another way to model the long-term evolution of a tidal basin is to formally integrate the mathematical equations of physical processes over the time and space domain. Since these equations are normally nonlinear, the closure terms should be involved. The closure terms have to be modeled in terms of large-scale dependent variables, empirically or based on a theoretical analysis of the relevant interaction processes (De Vriend 1996).

This approach is used by Krol (1990) to integrate a simple 1D morphological model of a tidal estuary. Later Schuttelaars and De Swart (1996) followed the same approach.

1.3   Goal of this study

Many efforts have been made during the last decades in the long-term morphological modeling of tidal basins. These studies are mainly based on the use of 'behavior oriented models'. This kind of models usually cannot describe the underlying processes and mechanisms of morphological evolutions; hence process-based models may provide better understanding of the morphological behavior of the tidal basins. On the other hand our ability of using process-based models for long-term simulations should also improve. These two goals go parallel to each other, therefore following the hypothesis that 'If you put enough of the essential physics into the model, the most important features of the morphological behavior will come out, even at longer time scales' (Roelvink, 1999) the main goal of this study is to :

Use process-based models to simulate the large scale morphological behavior of a tidal basin in order to gain more understanding of the physical processes governing the long-term morphological evolution of large tidal basins.

The goal of this study is chosen to challenge the ideas such as: Process-based models are designed to represent typical short- term processes and are validated against data concerning those processes; long-term developments are governed by other, more subtle processes, which can be dominated by the short-term 'noise', or, if already at short time-scales the process-based models are so sensitive to input parameters and numerical settings, does a long-term application of the same model make any sense? (NWO-ALW, 1999, sub-project 2, Wang et al. 2012)

1.4 Relevance

The main idea of this research was originally triggered by the gap in our understanding of tidal basins morphological behavior in the time scales between engineering time scales (short-term) and geological time scales (very long-term). Later it was felt that some "What if ..." questions, such as (in case of the Dutch Waddenzee) 'what would happen if the Afsluitdijk was not built?' need to be answered. Although the 'behavior oriented models' can give some answers about what would happen, a process-based model can shed some light on why it would happen. The knowledge which is gained about the processes and mechanisms of long-term morphological behavior of tidal basins provide us with the ability to predict the morphological changes of tidal basins better and can help the decision makers dealing with management of tidal basins to be able to foresee the consequences of their plans and decisions.

1.5 Research questions

Within this study, the following specific research questions are addressed :

To what extent does long-term process-based morphological modeling produce sensible results and which morphological features can be simulated by them?

Using process-based models for longer time scales is a new approach, which should be validated and examined carefully. Also since in tidal basin morphodynamics a wide variety of spatial and temporal scales are involved and various morphological features with different scales respond differently, temporally and spatially, to changes in forcing and human interventions, it is important to have an understanding of the morphological features that can be simulated in a long-term process-based morphological modeling.

What are the most important processes in long-term morphological modeling of tidal basins?

Most of the time using a process-based model means reducing the physical processes to be modeled ('Model Reduction') and/or schematizing the input to the model ('Input Reduction'). In any case it is important to know the most important processes related to the temporal and spatial scales of morphological modeling. Therefore it is essential to distinguish the processes which play the major role in the long-term morphological evolution of the tidal basins and the effect of reducing different processes from the model on the outcome of simulations.

What are the mechanisms governing the large scale morphological changes in tidal basins?

There are some morphological phenomena which can be observed from the historical data in tidal basins, such as the maximum depth of channels, channels migration, formation of new channels, evolution of ebb-tidal deltas, etc. In order to be able to predict the future behavior of a tidal basin it is necessary to understand the physical processes that are responsible for the development of these phenomena.

Can long-term process-based models be used to assess the effect of human interventions on the evolution of tidal basins?

The response of tidal basins such as Waddenzee to human interventions such as sea level rise, sea bed lowering (due to gas extraction) and closure of Zuiderzee is still quite uncertain. Time scale of these reactions and adaptations is in order of decades or centuries, therefore a long-term model can help in investigating these reactions.

1.6 Spectrum of the simulations and the structure of the study

In the first step of this study the performance of the process-based model in long-term simulations is investigated. Choosing the Dutch Waddenzee as the case study, a set of 2100 year long morphological simulations with the simplest possible setups is carried out. Adopting the realistic analogue approach the results of these simulations are compared qualitatively with the real morphology of the Waddenzee, as well as with equilibrium state empirical relationships. The reasonable results from these simulations and their good agreement with the empirical relations show that the process-based models can be used for longer durations (Chapter 2). In the next step the same principle is used to simulate the complete Dutch Waddenzee before the closure of Zuiderzee for 4000 years with a simple setup and the effect of applying the closure at different stages of the evolution of the tidal basins is investigated and compared with the existing conceptual models and hypotheses (Chapter 3). After this step some more complicated simulations with more processes are carried out. The first process which is added to the simulations is the sediment mixture. The effect of considering more realistic bed composition in the simulations is investigated by running 500 year long conceptual simulations as well as 75 years hindcast simulations. The outcome of the hindcast simulations is compared with the available data or in other words the process-based model is used adopting the virtual reality approach (Chapter 4). The next important process to be added to the simulations is the wave action, however the available knowledge and techniques about including wave in long-term morphological simulations is considerably limited. Therefore initially a conceptual model of a smaller tidal inlet system in Dutch Waddenzee (Ameland) is set up and the effect of wave or tide dominancy on the morphological evolution of tidal inlet systems is studied and the existing conceptual models are reproduced and explained by physical processes in the numerical model (Chapter 5). Finally in the last step, by means of 15 year long hindcast simulation of the Dutch Waddenzee the effect of different wave schematization methods for long-term process-based simulations is studied. Then choosing one of the methods, a complete 50 year long hindcast simulation focusing on the largest tidal inlet system in the Dutch Waddenzee (Texel/Marsdiep) is carried out. Overall the range of duration of the simulations in this study varies between 15 to 4000 years, clearly in cases with more processes the computational time per morphological year increases, therefore the simulations with more processes are carried out for shorter durations. Simulations with more processes are expected to reproduce the reality better; therefore the results of these simulations can be compared better with real data. On the other side of the spectrum are the models with longer durations and fewer processes, which are used to explain the hypotheses and conceptual model. Figure 1-2 shows the spectrum of all the simulations in this study.

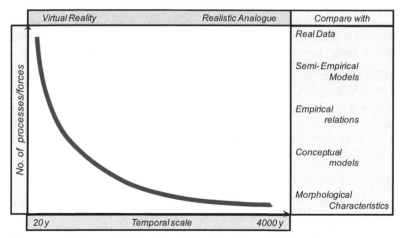

Figure 1 - 2 Spectrum of the simulations in this study

Each chapter of this dissertation is written as a standalone article, consequently the area of the case study and the description of the model which is used are repeated in each chapter but with more emphases on the concepts related to the topic of the each chapter.

CHAPTER 2

So far so good

Long-term process-based morphological modeling of
the Marsdiep tidal basin[1]

---

[1] A slightly modified version of this chapter is published as
Dastgheib, A., Roelvink, J.A., Wang, Z.B., 2008. Long-term process-based morphological modeling of the
Marsdiep tidal basin. Mar. Geol.. doi:10.1016/ j.margeo.2008.10.003

2.1   Introduction

Barrier islands, tidal inlets and tidal basins are found in many places along the coastlines in the world, and represent 10-13% of the world's continental coastlines (Schwartz, 1973). The flow through tidal inlets is basically driven by tidal oscillations, leading to periodic flooding and draining of the back barrier basin. Wind-driven currents may also contribute to this flow. The development and evolution of tidal basins is mainly due to the interaction between tidal currents, longshore currents, wave and river flow (if present); the geological features and overall geometry of the basin play their part as well. Tidal basins are very important since these are normally very rich ecosystems, hosting various valuable species. From the socio-economic point of view, tidal basins provide substantial opportunities for different local activities such as fishery, small industries and tourism; some medium size ports are also situated in tidal basins.

Tidal basins all around the world are subjected to different natural disturbances and human interventions, such as sea level rise, sand mining, dike constructions, coastal defense projects, land reclamations, basin closures, dredging etc. All these disturbances have some major effects on the morphology of the tidal basin and in turn, these affect the environmental and socio-economical value of the tidal basin. Large-scale human interventions can change the morphological behavior of the whole coastline rather than only the basin (e.g. the closure of Zuiderzee – the southern part of Dutch Waddenzee). Thus, long-term modeling of the morphological changes and evolution of tidal basins is needed to provide essential data for decision makers in coastal zone management.

The morphodynamic behavior of tidal basins is complex, mainly because of the wide variety of spatial and temporal scales involved. The whole basin system (mega-scale), different morphological elements (macro-scale), and various morphological features inside each element (meso-scale) respond differently, temporally and spatially, to changes in forcing (Scale classification from De Vriend 1991). These differences are to such an extent that the whole basin system may be in an equilibrium condition while there are large fluctuations within the elements.

In the last few decades, efforts have been made to identify equilibrium and stability of tidal inlets and to model morphological changes in tidal basins in different temporal and spatial scales using behavior based models (De Vriend et. al., 1993). These include empirical relationships, such as tidal prism-cross sectional area relationship, (e.g., O'Brien, 1931; Jarret, 1976) and closure criteria (e.g., Escoffier, 1940); and semi-empirical long-term models such as ASMITA (Stive et al. 1998, Stive and Wang, 2003). With the recent improvement in numerical process-based morphological models, they have been used to simulate the morphological evolution of tidal basins on different time scales (Wang et al. 1995, Marciano et al 2005, Van der Wegen and Roelvink, 2008). These studies show that process-based morphological models, describing flow field, resulting sediment transport and bottom changes perform well in complicated morphological situations in tidal basins, not only in short-term simulations but also in long-term ones.

2.2   Objectives

The aim of this study was to investigate the ability of morphological process-based models to simulate the evolution of the mega and macro-scale morphological features in a tidal basin on a very long time scale. In this study the tidal inlets in the Dutch Waddenzee were used as a case study and the morphological changes of tidal basins in this sea are modeled for a sufficiently long period for achieving equilibrium (~2000 years) in order to answer the following questions:

- Can a schematized long-term process-based morphological model predict a hypothetical mega-scale stable situation for a tidal basin (the Marsdiep basin, in this case) based on given constant boundary conditions?
- Is the result of schematized long-term morphological modeling of the Marsdiep basin consistent with empirical relations?
- What is the effect of nearby basins in the Waddenzee on the morphology of the Marsdiep basin?

2.3 Study area

The Waddenzee, located at the south east side of the North Sea, consists of 33 tidal inlets system along the approximately 500 km of the Netherlands, Germany and Denmark coastlines. The barrier islands of these tidal basin systems separate the largest tidal flat areas from the North Sea (Elias 2006). The part of the Waddenzee which is along the Netherlands coastline (Dutch Waddenzee) is shown in Figure 2-1. The ebb-tidal delta shoals in the Dutch Waddenzee are relatively large while they are associated with relatively narrow and deep channels; the back barrier basins of these tidal inlet systems consist of extensive systems of branching channels, tidal flats and salt marshes. The main area of interest in the current study is the Western part of the Dutch Waddenzee, especially the Marsdiep basin.

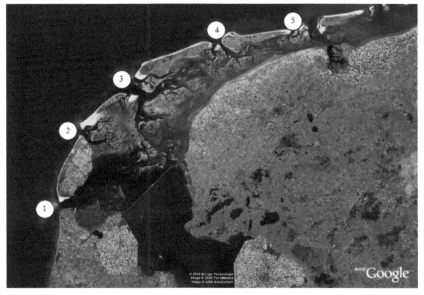

Figure 2 - 1 Satellite image of Dutch Waddenzee: 1- Texel-Marsdiep 2- Eierlandse Gat 3- Vlie 4- Amelander Zeegat 5- Friesche Zeegat

The Waddenzee is a young geological landscape, which has been subjected to numerous large or medium-scale human interventions such as closure of basins, land reclamation, coastal defense structures, sand nourishments etc. The human intervention which had the largest impact on the morphology of the Dutch Waddenzee is the closure of the south part of the basins, the Zuiderzee. Elias et al (2003) summarized the effect of this construction on the hydrodynamic and morphodynamic behavior of Waddenzee.

It has been shown that with regards to all the interventions and natural disturbances, the Waddenzee in its current situation is not in an equilibrium condition. Stive and Eysink (1989) note that the main cause of large and structural sand losses from the North-Holland coastline is the demand of sand in the Waddenzee tidal basins. Elias (2006) shows that the Marsdiep basin imports a large volume of sediment (3-5 million m^3) from the adjacent coast and ebb tidal delta every year. Based on theoretical knowledge and bathymetry data analysis, a conceptual model for the development of Waddenzee tidal basins is introduced by Elias et al (2003). This model describes the morphological development of Waddenzee in four different stages. In stage one, which is before human interventions, it is assumed that the whole system of Waddenzee is in a dynamic equilibrium. In this stage the characteristics of morphological elements of tidal basins can be described with empirical relations. This dynamic equilibrium was disturbed with the closure of the Zuiderzee in 1932. Stage two or 'adaptation period' is the period of large changes. In this stage, the natural behavior of the tidal basin systems is dominated by human interventions. Therefore, the empirical relations of equilibrium can not describe the morphological development of the tidal basin systems. This stage has a time scale in the order of several decades, and leads the system to a 'Near Equilibrium State'. In this stage (stage 3), the adaptation continues but on a long-term time-scale. Finally after centuries, the whole system will gain its new dynamic equilibrium state, clearly different from its original one (Stage 4).

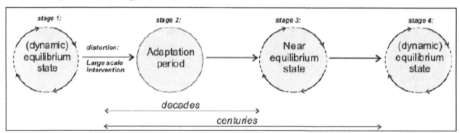

Figure 2 - 2 Conceptual model for Waddenzee tidal basins. (Elias et al. 2003)

It seems that the condition of Waddenzee now is somewhere at the end of stage two and beginning of stage three.

The Dutch Waddenzee is one of the best monitored coastal regions in the world. There are some depth measurements especially in the Marsdiep from the 16th century. Since 1987, Rijkswaterstaat (Directorate-General of Public Works and Water Management of The Netherlands) has frequently measured the bed level in the Waddenzee. The ebb-tidal deltas are measured every 3 years, while the basins are measured every 6 years. Rijkswaterstaat, based on the highest level of flat areas between the basins in the 1950's, has defined borders between different basins and the data for each basin is stored in a 20 x 20 m resolution database called 'Vaklodingen'. The available data before that time is less frequent and also less accurate; and is stored in a 250 x 250 m grid.

2.4   Model description and setup

2.4.1   Model description

The model which is used in this study is the 2DH version of the Delft3D model, described in Lesser et al. (2004) in detail. Basically, the governing equations of the same model are integrated over depth. The model uses a finite difference-scheme, which solves the momentum and continuity equations on a curvilinear grid with a robust drying and flooding scheme. For this exploratory study, the simplest possible physics (depth-averaged shallow water equations, simple transport formula) is applied. The sediment transport formula of Engelund-Hansen is used. As we are only interested in large scale development, the relaxation effect of suspended (sandy) sediment transport can be neglected, so no distinction needs to be made between bed-load and suspended transport.

$$S = S_b + S_s = \frac{0.05\alpha U^5}{\sqrt{g}C^3\Delta^2 D_{50}}$$

In which

U [m/sec]	: Magnitude of flow velocity
Δ [-]	: Relative density
C [m^{0,5} /sec]	: Chézy friction Coefficient
D$_{50}$ [m]	: Median grain size
α [-]	: Calibration coefficient (*O(1)*)

The approach adopted for morphological modeling in this study is called 'online approach' (Lesser et al., 2004, Roelvink, 2006). In this approach the flow, sediment transport and bed-level updating run with the same (small) time steps. Since the morphologic changes are calculated simultaneously with the other modules, coupling errors are minimized. But, as described in Lesser et al. (2004), because this approach does not consider the difference between the flow and morphological time step, a 'morphological factor' should be applied to increase the rate of depth changes by a constant factor (n) in *each* hydrodynamic time step. In this model even if a large value is chosen for *n*, the bed level changes are computed in much smaller time steps than in other approaches, e.g. tide averaging and rapid assessment morphology approach. The drying and wetting areas are also treated in a more straightforward way than e.g. in the classical tide averaged approaches (Roelvink et. al, 1994, Steijn et. al, 1996, Cayocca, 2001, Roelvink, 2006). Examples of the practical application of this approach can be found in Lesser et al. (2003, 2004). This method has also been used for detail event-scale modeling (Roelvink et al, 2003) for the case of breaching of a sand dam or narrow barrier island. For long-term morphological modeling of tidal basins and estuaries this method is used by Van der Wegen and Roelvink (2008) and Van der Wegen et. al (2008). The flow chart of the model is shown in Figure 2-3.

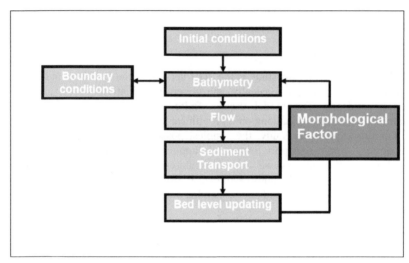

Figure 2 - 3 Model flowchart

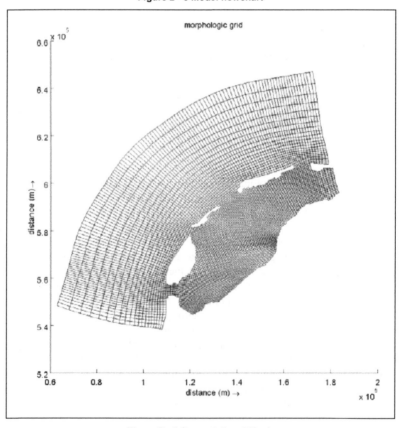

Figure 2 - 4 Computational Mesh

2.4.2   Grids

A local model for the Western Dutch Waddenzee (Figure 2-1) was set up. Although the main area of interest is Marsdiep, Eierlandse Gat and Vlie, the model was extended to the tidal divide between Amelander Zeegat and Friesche Zeegat to avoid boundary effects. Our interest was to set up a model with a reasonable computational time that can simulate long-term (~ 2000 years) morphological changes. The grid generated is a compromise between enough resolution in the inlets (at least 10 at the gorge) and having as few cells as possible (~ 7000 cells in total). The average spacing between grid lines inside the basins is about 350 m. The grid cells are smaller inside the basins and much bigger at the offshore boundary. The grid mesh covers only the area under the high water and the other parts of the barrier islands are excluded from the model, hence the sides of the inlets can not be eroded. Based on these considerations, the mesh shown in Figure 2-4 was generated for the study area.

2.4.3   Forcing

The main forces acting on a hydro-morphological model for coastal regions are tides, wind, waves and gravitational circulations. However, in this exploratory study, we focus on the effect of tidal forcing while ignoring other processes.

In order to determine the boundary conditions of this local model, a calibrated model for the tide in the North Sea, called 'ZUNO', is used.

The 'ZUNO' model is based on the 'Zuidelijke Noordzee model' from the Dutch Ministry of Public Works and was constructed by Delft Hydraulics. A detailed description of the calibration and validation of the model can be found in Roelvink et al. (2001).

The 'ZUNO' model has approximately 20,000 computational grids. In the coastal zone the grid sizes are approximately 1.5 km alongshore and 400m cross-shore. The model is forced by the boundary conditions on two open boundaries. The southern boundary is situated south of the Strait of Dover and the northern boundary lies between Scotland and the north of Denmark. At these boundaries, water levels are specified as astronomical components, amplitudes and phases of tidal constituents. The calibration of the model was based on comparing the water levels obtained from the model and observations from 47 locations.

Following Van de Kreeke and Robaczewska (1993), the spring neap cycle is ignored and the dominant forcing by M2 and over-tides is considered. The ZUNO model was therefore run with the forcing boundary conditions of M2, M4 and M6 until a periodic solution was reached. During this run, tidal level variations at the boundaries of the local model were recorded.

From the results of the ZUNO model, recorded tidal variations at local model boundaries were analyzed and M2, M4 and M6 were extracted for these boundaries. These components were used to form boundary conditions for the local model.

The local model has 3 open boundaries: one at the sea side and two lateral boundaries. The sea side boundary is chosen to be a water level boundary, while the lateral boundaries are Neumann boundaries, where the alongshore water level gradient is prescribed (Roelvink and Walstra, 2004). These boundary conditions allow the cross-shore profile of alongshore velocity and water level to develop without disturbances.

2.4.4   Initial bathymetry

In these simulations the geological constraints and variation of sediment grain size are neglected and a uniform grain size is used for the bed material. The other parameters which can affect the hypothetical equilibrium condition of tidal basins in this process-based modeling approach are the sediment thickness and initial bathymetry. The initial bathymetry will affect the evolution of the tidal basins because it affects the competition between tidal basins. If a tidal basin initially has larger area and deeper channels (like Marsdiep), it will keep its dominant role. In addition to the real bathymetry, two other types of schematized bathymetries are also used as initial bathymetry. In order to calculate the sediment thickness in each bathymetry, an arbitrary level was chosen and everything above that level was assumed to be (erodible) sediment. In the model there is no sink or source of sediment; therefore the total amount of sediment during the simulation is constant. It should be mentioned that the amount of sediment exchanged through the lateral open boundaries is negligible.

Real Bathymetry

For runs with real bathymetry, data from 1998 was used. This data was projected on the grid using triangular interpolation. This bathymetry is shown in Figure 2-5.

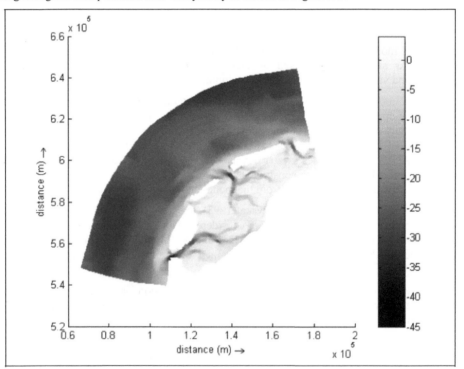

Figure 2 - 5 Bathymetry of 1998 projected on the grids

Flat bathymetry

An interesting way to model a tidal basin is to use a flat bathymetry inside the basin without any kind of ebb-tidal delta outside the inlet, to allow the model to show the mechanism of building and changing of the ebb-tidal delta outside the basin and the channel and shoal patterns inside. So it was decided to make schematized bathymetries with flat bed inside the Waddenzee. For this purpose the following steps were taken:

- Inside the Waddenzee the bottom was assumed to be flat, including at the inlets
- No ebb-tidal delta or channels were included in the bathymetry outside the basins
- The slope of the coastal shelf was made uniform
- The offshore side of the model was assumed to be flat

A sample of this bathymetry is shown in Figure 2-6.

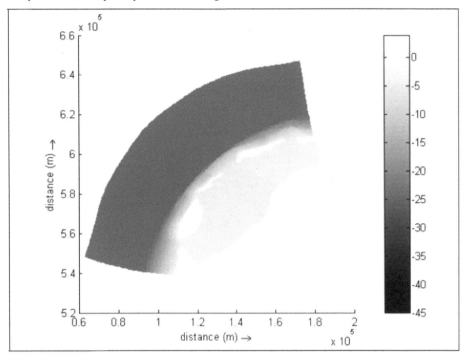

Figure 2 - 6 A sample of schematized bathymetry with flat bed level inside the Waddenzee

To determine the depth of the schematized flat Waddenzee, an analysis on the availability and distribution of the sediment inside the basins was carried out and, based on different criteria, different depths were chosen:

- Depth = 3.62: The volume of sediment inside the basins is equal to the combined volume of the three basins plus the volume of sediment in the ebb-tidal deltas.
- Depth = 4.54: The volume of sediment inside the basins is equal to that of the real bathymetry

- Depth = 5.02: The volume of sediment inside the Marsdiep basin is equal to that the real bathymetry plus the amount of sediment in its own ebb-tidal deltas

4.4.3. Sloping bathymetry

In recent studies, sometimes a sloping bathymetry toward the inlet is used as the initial bathymetry while attempting to model the morphological evolution of tidal basins with process-based models (e.g. Wang et al, 1995 Marciano et al, 2005). Similarly in this study, a schematized sloping bathymetry was made for the Waddenzee. The procedure of this schematization is as follows.

The tidal basins are separated based on the borders defined by Rijkswaterstaat in the 'Vaklodingen' database.

In each basin, the depth of grid points is determined as a function of the distance of grid point from the center point of the inlet at the basin side. This function is provided by fitting curves on the data from the measured bathymetry of 1998.

- The amount of sediment of each ebb-tidal delta is distributed uniformly inside the corresponding basin.
- The slope of the coastal shelf was made uniform
- The offshore side of the model was assumed to be flat

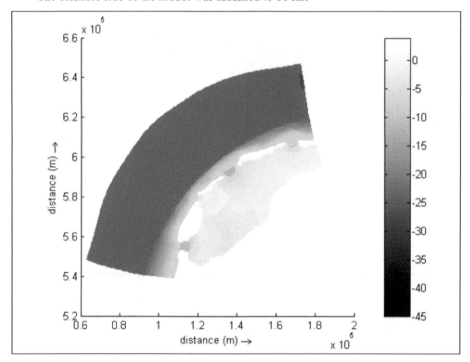

Figure 2 - 7 Sloping bathymetry

2.4.5   Different Runs

Considering different initial bathymetry conditions, 5 different simulations were carried out, one with the real bathymetry, 3 with flat bathymetry with the depths of 3.62, 4.54 and 5.02 m, and one with sloping bathymetry .

The bed material in all cases consists of uniform sand with D50 = 200 μm. For bottom roughness a Chezy value of 65 $m^{1/2}/s$ is used.

To choose the morphological factor, reference is made to Van der Wegen and Roelvink (2008). Their study shows that in long-term simulations of tidal basin with tidal forces, the main morphological characteristics of the basin are maintained if high values of morphological factor (up to 400) are used. In this study, the morphological factor of 300 is used and by running the model for 7 years of hydrodynamic time, 2100 years of morphological time is simulated. This period is expected to be sufficiently long for the system to adjust to its morphological equilibrium (Stive and Wang, 2003).

2.5   Model results and discussion

The result of the model in this study is described in three parts: the result of the model for Marsdiep Basin is discussed; the effect of different basins on each other is presented; and finally, the effect of historical background (initial bathymetry in the model) on the evolution of the basins in the multi-inlet tidal system of the Dutch Waddenzee is discussed.

2.5.1   Model result for Marsdiep basin

Morphological evolution

Starting with the schematized bathymetry inside the basin, the model shows the evolution of the ebb-tidal delta and channel and shoal patterns inside the basins. For example, in the run with a flat bathymetry with a depth of 4.54 m inside the basins, during the first 100 years of modeling the ebb-tidal delta in front of the Marsdiep inlet is formed, and also the main entrance channel is developed during this period (Figure 2-8). This is followed by the evolution of the main channel and shoal patterns inside the Marsdiep basin. After almost 400 years, the main channel and shoal pattern inside the Marsdiep basin is almost defined; however, the dynamic behavior of this pattern is obvious (Figure 2-8). The ebb-tidal delta migrates towards the inlet as well as northwards. The main channel inside the basin clearly stretches eastwards and does not change its orientation during this time while other channels show a fractal dynamic behavior till the basin reaches its relatively stable channels with three channels originating from the entrance channel. The same phenomena are observed in all the simulations with flat and sloping bathymetries.

In the simulations, the ebb-tidal delta is generated in the proper location with the asymmetry toward the direction of tide propagation. Also the north-eastward direction of the main channel inside the basin is in good agreement with the bathymetry of Marsdiep in 1998. However, the smaller scale features are not in agreement with reality. The maximum depth of the inlet gorge in the model is more than the actual value, which may be due to neglecting wave forcing or because of our simplified bed composition, which does not take into account geological constraints or armoring. Therefore, the shape of the ebb-tidal delta especially in the outer boundary is different from reality. Also, the coarse grid used in this model limits the width of the channels and shoals in the basin to an average of 500 m and channels less than this width cannot be generated in the model.

Another difference is the channel pattern. The simulated channel pattern is more or less fractal, whereas in reality the main channel pattern is more meandering than branching. This difference is mainly due to the morphological background of the Marsdiep. In reality, the Marsdiep basin including the Zuiderzee, used to be a long basin where the dominant channel pattern is meandering. After the closure of the Zuiderzee, this kind of pattern is still obvious in Marsdiep main channel. But in the model, Marsdiep is simulated after the closure of the Zuiderzee. A tidal basin with this dimension is not a long basin and the main channel pattern in such a basin is more fractal than meandering as shown in the results of the model (see section 5.3). Still, the orientation and location of the main channel are more or less in agreement with the present situation..

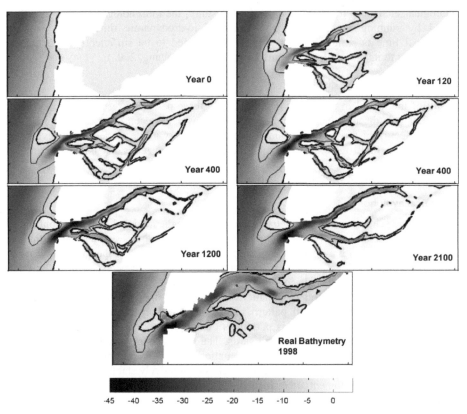

Figure 2 - 8 Simulated Evolution of Marsdiep Basin bathymetry and its ebb-tidal delta using Delft3D model from a flat bathymetry inside the basin, in morphological years of 0 (initial condition), 120, 400, 800, 1200, and 2100, compared to real bathymetry of 1998 (Contour lines are presented for -5 and -10 m depth)

Basin Characteristics

In the literature, characteristics such as relative flat area, flat height, tidal prism, etc are defined for a tidal basin and most of the empirical equilibrium relations are based on these characteristics. In this section, the sediment balance for the Marsdiep basin and a comparison of the model results to empirical relationships are presented.

Sediment balance

One of the most important characteristics of a tidal basin is the amount of sediment which is imported to or exported from it. As mentioned before, at present the Marsdiep basin imports a large amount of sediment (3-5 million m³ per year). The simulated change of the sediment volume in the Marsdiep basin (including the inlets) is calculated based on the sedimentation and erosion in the basin at each time step (Figure 2-9). This shows that in the cases with flat initial bathymetry, the Marsdiep basin first exports some sediment to form its ebb-tidal delta and then imports a large amount of sediment from the delta and the adjacent coast line. In these cases, the rate of sediment transport to the Marsdiep basin even after 2100 years of morphological modeling cannot be neglected, although this rate decreases throughout the simulation time. It is also clear that the final condition of the basin is highly dependent on the initial condition in the model. Simulation with the real initial bathymetry shows that Marsdiep in this model imports about 400 million m³ during 2100 years of simulation. This rate is not constant during the 2100 year period. The main portion of the sediment import takes place in the first 300 years. During the first 300 years about 300 million m³ of sand enters the Marsdiep with a maximum rate of about 3 million m³ per year in the first 40 years. In the last years of modeling, this rate reduces to only 30 thousand m³ per year. So it seems reasonable to claim that from the point of view of sediment exchange, the model reaches a stable condition.

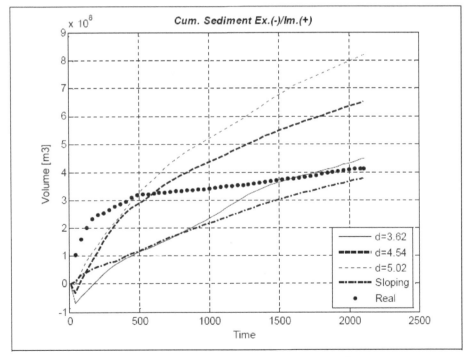

Figure 2 - 9 Sediment exchange in Marsdiep Basin in simulations with different initial bathymetry

Relative inter-tidal flat area

The inter-tidal flat area is defined as the area between MLW and MHW. In literature there are suggestions for flat areas in the equilibrium condition. De Vriend et al (1989) showed a general relation between the flat area and the total area of the basin. Renger and Partenscky (1974) worked on the same form of relation for inlets in the German Bight. Later, Eysink (1991) re-wrote their relation (showed in dashed solid in Figure 2-11).

The relative flat area (A_f/A_b) during the simulation time for different initial conditions is shown in Figure 2-10 and compared with the value based on the Renger and Partenscky (1974) observations in the German Bight inlets. It shows that the value of A_f/A_b also tends to a stable value. But this stable value is also dependent on the initial condition.

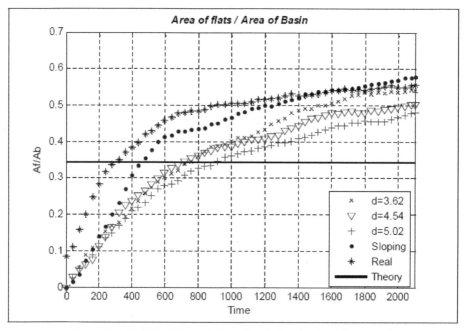

Figure 2 - 10 Relative flat area in Marsdiep during the simulation period

Eysink (1991) used the same idea (A_f/A_b as a function of A_b) to analyze the available data in tidal inlets and estuaries in The Netherlands. The outcome of his analyses for Waddenzee is presented in Figure 2-11 with solid lines. A_f/A_b during the simulation time is also plotted in this figure. It shows that the results of the simulation for this parameter, for all initial conditions, are in the range that Eysink (1991) suggested for Waddenzee.

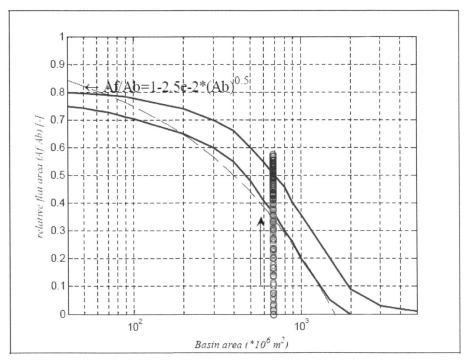

Figure 2 - 11 Relative flat area in Marsdiep (Circles: Result of modeling during the simulation time; dash line: the relation based on Renger and Partenscky (1974) observation; solid lines: Eysink (1991) observation for Waddenzee.

Height of flats

Eysink (1990) claims that one of the first parameters that aims for equilibrium in a relatively short time is the height of flats, which is related to the tidal amplitude. Height of flats, which is usually used in the equilibrium situation, is defined as the average height of the flat areas calculated by the following relation:

$$h_f = \frac{V_f}{A_f}$$

In which,

$A_f \, [\text{m}^2]$: Flat Area at MLW

$V_f \, [\text{m}^2]$: Volume of flats, i.e. volume of sediment in the region between LW and HW

$h_f \, [\text{-}]$: Height of flats

To check the Eysink's hypothesis in the results of process-based modeling, the development of flats in the Marsdiep basin is shown in the Figure 2-12.

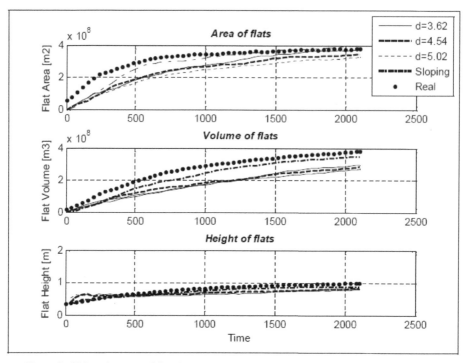

Figure 2 - 12 Development of flat characteristics in Marsdiep from different initial conditions

Figure 2-12 shows that flat characteristics tend towards equilibrium values but the flat height is not adjusted as fast as Eysink (1990) claims. It is also shown that equilibrium values for volume of flats and area of flats are dependent on the initial condition, while the flat height is almost the same for different initial bathymetry. The initial sloping bathymetry forced the model to develop more flat volume and also more flat area. So it can be concluded that the longitudinal distribution of the sediment in initial bathymetry also affects the results for flat characteristics. The final height of flats in all the simulations with flat initial bathymetries is almost the same but far from the equilibrium value suggested by Eysink (1990), which is around 0.4 m in the case of Marsdiep basin. The main reasons for this difference are probably the lack of wave stirring and the absence of horizontal sediment gradation, which leads to deeper channels and higher flats.

Ebb and Flood Dominance

Speer and Aubrey (1985) used a 1D numerical model to study the influence of geometry and bathymetry on tidal propagation of short, friction-dominated and well-mixed estuaries. They suggested that two non-dimensional parameters can be used to characterize the tidal basins into ebb or flood dominant. The first one is a/h, the ratio of the tidal amplitude and the depth of the channel with respect to MSL, which shows the relative shallowness of the estuary. The second parameter is the ratio of the volume of inter-tidal storage and channel volume (V_S/V_C). Larger values of a/h (shallower basin) means longer ebb duration (due to larger effect of friction and different wave propagation velocity), while increase in inter-tidal storage will decrease the flood propagation and duration. Later, Friedrichs and Aubrey (1988) confirmed the Speer model against measured data along the Atlantic coast of the United

States. Speer et al. (1991) translated the Friedrichs & Aubrey (1988) results into a graph (Figure 2-13) which distinguishes the flood or ebb-dominance. Donkers (1998) suggests that if duration of flood and ebb are equal, the equilibrium of the longitudinal bed profile of tidal basin is reached, which means that the border between the two regions in the graph would represent equilibrium conditions of the basin.

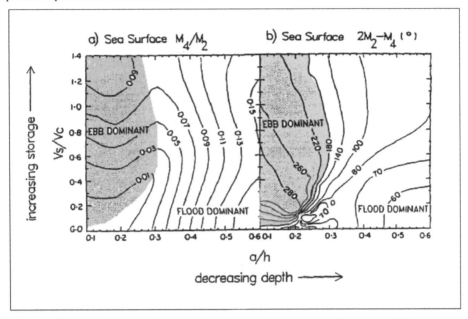

Figure 2 - 13 Diagram Based on Friedrischs & Aubrey models (Speer et al., 1991)

In this study, the Friedrichs & Aubrey graph is used as an indication for flood or ebb dominance. It should be mentioned that Friedrischs & Aubrey used a highly schematized model, assuming a constant and fixed longitudinal geometry, constant offshore forcing, and basin length of 7 km (Short basin) which are in contrast to the model used in the current study. Recently Van der Wegen and Roelvink (2008) used the same graph to distinguish between ebb and flood dominance in an 80 km long tidal basin model. Provided that the basin evolved towards a state of 90 degree phase lag between velocities and water levels, the graph indeed suggested equilibrium conditions.

In this study, from the results of the model for Marsdiep tidal basin, a/h is calculated by dividing the tidal amplitude by the average channel depth and plotted against ratio of the volume of inter-tidal storage and channel volume with Friedrichs & Aubrey graph (Figure 2-14). This graph shows that initially the basin is flood dominant and in all the simulations, development of the basin is towards the equilibrium line. This development is faster in the early years of modeling. This can be explained by the decreasing sediment import to the basin. The basin initially imports more sediment when it is more flood dominant. When the basin condition is near the equilibrium line, it begins to scatter and develop almost parallel to the line.

This evolution is due to changes in 3 different characteristics of the basin: inter-tidal storage (Vs), channel volume (Vc) and the average depth of channel (h). These changes in the simulation with the real initial bathymetry are discussed in the following paragraph.

In initial bathymetries of the simulations, even the real bathymetry of Marsdiep, the area of flats higher than mean low water is zero or very small, therefore Vs is very small or zero. During the first 700 years, the area of flats increases for the simulation with real initial bathymetry. The sediment needed to produce the flat areas is supplied from both imported sediment and deepening the channels. During this period, channels become narrower and deeper. Because the rate of the narrowing of the channels is higher than the deepening, the channel volume decreases in this period, so the ratio Vs/Vc increases. This trend continues till Vs reaches its maximum value. This maximum value is due to decline of the rate of expansion of flat areas. Meanwhile, the imported sediment slowly decreases the volume of inter-tidal storages. On the X-axis, deepening of the channels during the simulation period decreases the value of a/h, which happens at a higher rate in the beginning of simulations.

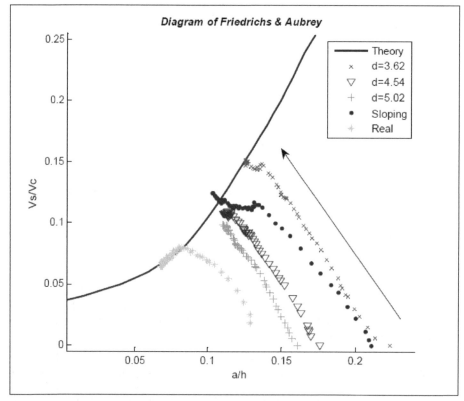

Figure 2 - 14 Friedrichs & Aubrey diagram for modeled Marsdiep with different initial condition, the arrow shows the direction of changes with time

2.5.2   The boundary between different basins

The interaction between the adjacent tidal basins in a multi-inlet tidal basin can be interpreted as changing of the boundaries of the area of influence of each tidal inlet or, in other words, the boundaries of each tidal basin. Looking at the result of the current long-term simulations, especially the morphological development in simulations with real initial bathymetry, it seems that the boundaries which are defined by Rijkswaterstaat in the 'Vaklodingen' database for different basins (based on the tidal divides in the 1950's) in the Western Waddenzee may not be a good choice over long periods and these boundaries or tidal divides are subjected to some changes. Also, the important role of these boundaries in analyses is mentioned in other studies in the same region, but to the best of our knowledge, no extensive study has yet been done on this issue and most of the investigations are based on predefined boundaries. Simulated changes of the boundaries of tidal basins are briefly discussed here.

To study the change of the tidal divides, results of one simulation with real initial bathymetry are used and the resulting bathymetry of morphological modeling is extracted from the results after 0, 200, 700, 1200, and 2100 years. Then, to provide the corresponding flow field for all these years, 5 hydrodynamic simulations for 5 tidal cycles are carried out using the extracted bathymetries. Having the bathymetry and flow field in each case, it is possible to define the boundaries between tidal basins in the model of the Western Dutch Waddenzee.

The physical boundary between the tidal basins is the tidal divide, but to find it in the results of the simulations this definition should be translated to flow field or bathymetry characteristics. In this study, it is suggested that the line of minimum standard deviation over a tidal cycle of (depth averaged) velocities is the tidal divide. This definition is applied to the results of simulation in different years and the boundaries of the basins are defined. The boundaries of the basins thus found are plotted in Figure 2-15 for different years.

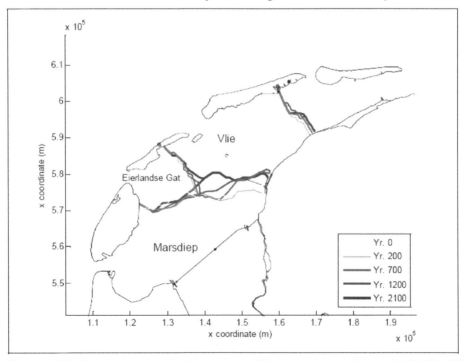

Figure 2 - 15 The boundary between the tidal basins in different morphological years: 0, 200, 700, 1200, and 2100

The results clearly show that the boundaries of the basins are stretching eastward; Marsdiep is gaining area from Vlie, while Vlie in turn is extending its boundary into the Amelander Inlet. But in the case of Eierlandse Gat, the boundary of this basin does not expand; this boundary only rotates toward the east. In a parallel study on the measured data during 1926-2006, the same behavior is observed (Van Geer, 2007).

2.5.3 Effect of initial bathymetry

The effect of initial bathymetry (or historical background) of the basins on the evolution and interaction between basins is also investigated in this study. Figure 2-16 shows the final bathymetry of the Dutch Waddenzee after a 2100 year simulation with different initial bathymetries compared to the real bathymetry of 1998.

As it is shown in panel (f) of Figure 2-16, in reality the Eierlandse Gat basin is much smaller than Marsdiep and Vlie, and during the evolution of the Waddenzee this basin is dominated by the two other basins. But the model results show that if the evolution of tidal basins begins with the same initial condition (panel a, b, and c), the differences between the sizes of the basins are marginal and all three inlets grow to a more or less similar state. In the simulations with initial sloping and real bathymetry, the Eierlandse Gat basin is much smaller than Marsdiep and Vlie. This shows that the amount of sediment in the basins as well as the historical form of the channels and shoals in tidal basins can affect the evolution of the basins. Moreover, the effect of the historical background of the Waddenzee is pronounced in this comparison. As is clear, the simulated channel and shoal pattern is more fractal in the Marsdiep and Vlie basins in simulations with schematized initial bathymetries, rather than the ones with real initial bathymetry.

2.6 Conclusion

In this study it is shown that a process-based model can be used to simulate long-term morphological changes in tidal basins and produces reasonable results. The process-based model which is used in this study does not simulate one single *mega-scale* stable (equilibrium) condition in the Western Waddenzee including the Marsdiep for all initial conditions for the *duration of the simulations*, but with each initial condition in many aspects such as sediment exchange and some basic characteristics of tidal basins, a *mega-scale* stable (equilibrium) condition is simulated, which is dependent not only on the given forcing boundary condition but also on the initial condition.

It is shown in the analyses that the effect of adjacent basins on each other can be interpreted as changes in the boundaries of basins. From this point of view, the Marsdiep basin is stretching eastward and gaining some area from the Vlie, while the Vlie is also expanding towards east. Generally the interaction between different tidal basins in a multi-inlet tidal system such as Dutch Waddenzee plays a significant role in determining the basin characteristics.

In this study it is shown that the results of this process-based model follow the empirical equilibrium equations for flat characteristics and relative flat area qualitatively. Also, it is shown that the Marsdiep tidal basin, during the simulation period, becomes less flood-dominant.

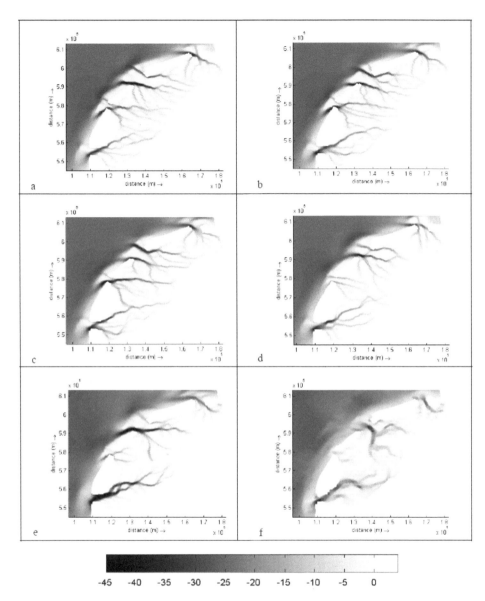

Figure 2 - 16 The resulting bathymetry of Waddenzee after 2100 Morphological years simulation from different initial bathymetries : a) Flat bathymetry, d= 3.62m b) Flat bathymetry, d= 4.54m c) Flat bathymetry, d= 5.02 d) sloping bathymetry e) real bathymetry, compared to bathymetry of 1998 (f)

CHAPTER 3

Turning the tide

Long-term morphodynamic effects of closure dams on tidal basins

3.1   Introduction

Tidal basins are present in many places along the coastline of the world. Long-term morphological evolution of these basins is of a great importance from both scientific and socio-economic points of view. Channels in the basins provide natural, deep-water access to ports, and intertidal flats often form unique ecosystems (Allen, 2000). Man has started to influence the natural evolution of tidal systems since a thousand years ago by land reclamation, dredging and closing of the basins. Closure dams in lowland countries such as The Netherlands are traditionally used to protect tidal inlets from occasional storm surge events, and/or to provide possibilities to reclaim new land from the tidal basin area. Generally it is assumed that the morphology of long existing tidal inlet systems has reached an equilibrium state, which can be described by empirical relations between morphology and tidal motion (Escoffier 1940, O'Brien 1969, Walton and Adams 1976, etc.). Constructing a closure dam often causes an instantaneous change in tidal wave propagation and flow field in the basin which disturbs the existing equilibrium, and triggers extensive morphological changes in the adjacent tidal basin. These morphological changes may continue for decades or centuries before the whole system reaches a new equilibrium state, which can be different from the initial equilibrium before the closure (Figure 3-1). Kagtwijk et al. (2004) and Van de Kreeke (2006) used different aggregate models to explain the long-term morphological response of tidal inlet systems to reduction of the tidal basin area by constructing a closure dam. In recent years advances in the knowledge of numerical modeling of the physical processes together with technological developments, made it possible to use process-based models for long-term morphological simulations. Therefore now it is possible to study the morphological behavior of complex coastal systems such as tidal basins using this type of modeling technique. (Hibma et al 2003, Van Leeuwen at al. 2003, Van der Wegen and Roelvink, 2008, Van der Wegen et. al. 2008, Dissanayake et. al. 2009). Using a process-based model and considering a realistic analogue approach (Roelvink and Reniers, 2012) enable us to select a process that induces morphological changes of tidal basins and provide more insight into the effect of large human interventions such as closure dams on that specific process and consequently on the morphological behavior of the tidal basin.

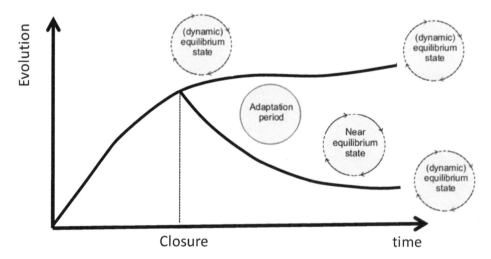

Figure 3 - 1 Schematic description of the effect of human intervention on the morphological evolution of tidal inlet systems from Elias et al (2003)

3.2 Aim of Study

The main goal of this study is to investigate the characteristics and time scale of the effects of the construction of a closure dam on the morphodynamics of tidal inlet systems. To achieve this goal we have used a numerical process-based model and have chosen the Dutch Waddenzee tidal basins as the case study.

3.3 Study area

The Dutch Waddenzee is the western part of the Waddenzee on the south-east side of the North Sea along the Netherlands coastline. This part of the Waddenzee consists of five tidal inlets connecting the North Sea to the basins behind the barrier islands (Figure 3-2). Before 1932, the Texel and Vlie inlets covered the south-western part of the Waddenzee and the former Zuiderzee. In 1932 the south part of the basins (Zuiderzee) was closed by a 32 km long dike known as the Afsluitdijk. After the construction of the Afsluitdijk, the area covered by the tidal basins in the Western Dutch Waddenzee decreased from around 4000 km^2 to 712 km^2 and as it was predicted the tidal range at the inlet increased by about 15 % and the tidal prism by about 20% (Elias et al. 2003). This dramatic change is the main reason for the ongoing morphological changes in the present day Dutch Waddenzee. Using field based evidence, Elias et al. (2003) extensively described the hydrodynamic and morphodynamic changes of the Texel inlet and the Marsdiep basin in the Dutch Waddenzee due to this closure, and suggested a conceptual model that explains the impact of the closure on the tidal basins. It is observed that at present the Texel and Vlie inlets are separated by a tidal divide resulting in minor exchange between the two basins (Ridderinkhof, 1988). Figure 3-3 shows an impression of tidal wave propagation in the Dutch Waddenzee prior to and after the closure (Elias et al. 2003).

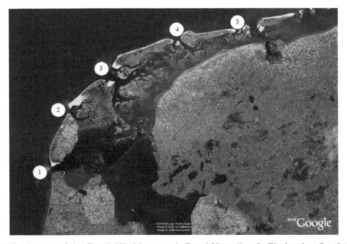

Figure 3 - 2 Satellite image of the Dutch Waddenzee: 1- Texel-Marsdiep 2- Eierlandse Gat 3- Vlie 4-Amelander Zeegat 5- Frisian inlet (image courtesy Google Earth).

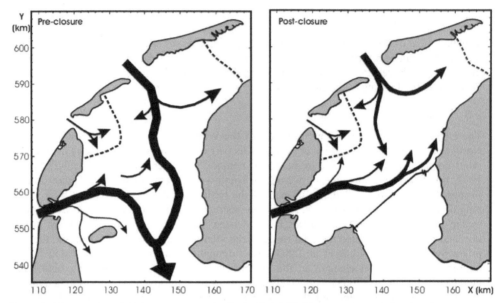

Figure 3 - 3 Impression of tidal wave propagation in the Dutch Waddenzee prior and after the closure (Elias et al. 2003)

Elias et al (2003) compared the bathymetry of the basin just after the closure and the bathymetry 65 years after the closure. They illustrated the drastic changes in the morphology of Texel inlet and Marsdiep basin adapting to the new conditions: large sedimentation in the terminal parts of the channels, lateral channel migration towards the east, separation of the Texel and Vlie sub-basins by a tidal divide, reduction in the volume of ebb-tidal delta and sediment imports into the Waddenzee basin. Elias et al. (2004) used a process-based model and simulated the hydrodynamic and sediment transport of the Texel inlet and Marsdiep basin in various stages of the morphological adaptation and validated the conceptual observation-based ideas of the morphological adaptation.

3.4 Model Description

The model which is used in this study is the 2DH version of the Delft3D model of which a complete description is given by Lesser et al (2004). Basically the governing equations of the same model are integrated over depth. This model is a finite difference-scheme model which solves the momentum and continuity equations on a curvilinear grid with a robust drying and flooding scheme . For this study, the simplest possible physics (depth-averaged shallow water equations, simple transport formula) are applied. In this study the empirical relation of Engelund and Hansen (1967) is used for sediment transport.

$$S = S_b + S_s = \frac{0.05\alpha U^5}{\sqrt{g}C^3\Delta^2 D_{50}}$$

In which

U	[m/sec]	: Magnitude of flow velocity
Δ	[-]	: Relative density
C	[m^{0,5} /sec]	: Chézy friction Coefficient

D50   [m]   : Median grain size

α   [-]   : Calibration coefficient (O(1))

Following Roelvink (2006) we have used the so called online approach. In this approach the flow, sediment transport and bed-level updating run with the same (small) time steps (Lesser et al, 2004, Roelvink, 2006). Since the morphologic changes are calculated simultaneously with the other modules the coupling errors are minimized. But since this approach does not consider the difference between the flow and morphological time step, a 'morphological factor' to increase the depth changes rate by a constant factor (n) should be applied. So after a simulation of one tidal cycle in fact the morphological changes in n tidal cycles are modeled.

3.5 Model Setup

3.5.1 Model Domain

A model for the Dutch Waddenzee before closure is set up. The land boundary of the model is determined using historical maps together with the borders of Dutch new municipalities. In this study we need to approach an equilibrium condition therefore we have to simulate for a very long-term (~ 4000 years) morphological changes, which means that the computational time is a matter of concern. Therefore the grid we generated is a compromise between enough resolution in the inlets (at least 10 grid cells at the gorge) and having overall as few cells as possible to obtain a reasonable computational time. In this study the average spacing between grid lines inside the basins is about 350 m. The grid cells are smaller inside the basins and they are much bigger at the offshore boundary. The computational grid covers only the area under the mean high water line and the other parts of the barrier islands are excluded from the model. Figure 3-4 shows the domain and the computational grid of the model without closure. Since the main interacting inlets are Texel and Vlie and the Eierlandse Gat inlet is confined naturally due to the geological setting, we have closed off the tidal basin of this inlet from the rest of the basins.

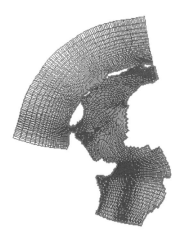 

Figure 3 - 4 The model domain and the computational grid of the model

3.5.2   Forcing

Similar to Chapter 2, the only hydrodynamic process which is used in the simulations in this study is the tidal forcing. Other forcing mechanisms such as wave and wind climate are not included in the simulations. In order to determine the boundary conditions for this local model, a calibrated model for the vertical tide in the North Sea, called 'ZUNO' (Roelvink et al., 2001), is used.

Referring to Van de Kreeke and Robaczewska (1993) and Latteux (1995), we neglect the spring neap cycle and consider M2 and overtides as the morphological tide. Therefore the ZUNO model is run with the forcing boundary conditions of M2, M4 and M6 until a periodic solution was reached. During this run the tidal level variations at the boundaries of the local model are recorded.

From the results of ZUNO model the recorded tidal variations at the boundaries of the local model are analyzed and the M2, M4 and M6 components are extracted for these boundaries using Fourier analysis. These resulting amplitudes and phases are used as boundary conditions for the local Waddenzee model.

The boundaries of the local model consist of 3 segments: one boundary at the sea side and two other lateral boundaries. The sea side boundary is chosen to be a water level boundary, while the lateral boundaries are Neumann boundaries, where the alongshore water level gradient is prescribed (Roelvink and Walstra, 2004). The same boundary conditions have been used for the models both with and without the closure dam. Therefore it is assumed that the offshore boundary of the model is far enough not to be affected by the closure.

3.5.3   Initial Bathymetry

The initial bathymetry which is used in the simulations of this study is a flat bathymetry inside the basins without any kind of ebb-tidal delta outside the inlets. Therefore, based on the applied forcing and the available sediment volume, the model simulates the mechanisms of building and changing the ebb-tidal delta outside the basin and channel and shoal patterns inside the basin. In this case the uniform depth of this flat bathymetry is equal to average depth of the basins (depth of 4.54 m below mean sea level) .

3.5.4   Scenario of simulations

In this study a morphological simulation is carried out for 4000 years including the whole domain (Waddenzee together with the Zuiderzee). To achieve a reasonable computational time a morphological factor should be applied, Ranasinghe et al. (2011) suggest that the value for morphological factor should not be more than 100, however . Several other studies have simply concluded that MFs as high as 400 are capable of producing realistic results based on qualitative comparisons with measured bathymetry (Dissanayake et al., 2009; Van der Wegen et al., 2008; Van der Wegen and Roelvink, 2008). In this study we have used the value of 300 for the morphological factor. Using this procedure , the morphological "equilibrium" state which is reached in this simulation is only due to the tidal forcing. To investigate the effect of the closure on the morphology of the basins, in subsequent runs we have applied the closure after 1000, 2000, 3000 and 4000 years and continued the simulation including the closure dam (the same model domain excluding the Zuiderzee) for another 2000 years. Figure 3-5 shows the scheme of these simulations.

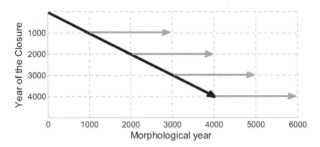

Figure 3 - 5 Scheme of the simulations, arrows show different simulations

3.6   Results and Discussion

3.6.1   Morphological changes

Figure 3-6 shows the 4000 years morphological evolution without the closure and with the closure at simulation year 2000 as well as the resulting bathymetry of the whole domain in simulation year 2000. It is clear that in the first 2000 years the channels in Vlie and especially Marsdiep are penetrating into the Zuiderzee and this procedure continues for the next 2000 years if there is no closure dam, the main channels of Marsdiep are connected to the Zuiderzee and have a south-easterly orientation. However in the case in which the closure is applied after 2000 morphological years, the main channel in Marsdiep reorientates and migrates towards the east and the connection between Vlie and Marsdiep is cut. The areas closer to the closure dam get silted up and some flat areas develop in that area. The main channels in the south-west of the Waddenzee lose their importance. In Figure 3-7 we have shown 2 cross-sections for the channels: one in the Marsdiep basin and one very close to the location of the closure dam. After the closure the most western channels in the Marsdiep basin get shallower. Without the closure those channels serve as the main channels of the basin. Further, due to the closure the eastern channel, the main channel of the Marsdiep migrates towards the east (Figure 3-7a). At the cross-section close to the location of the closure dam we can observe that the main channel connecting the Waddenzee to the Zuiderzee is completely silted up, however in the case without the closure dam this channel would increase in depth and play the main role in the exchange of water and sediment between the Waddenzee and the Zuiderzee. At the east side of this cross-section two small channels are generated as a result of the reorientation of the channels of the Marsdiep towards the east. There is also sedimentation taking place due to the closure at this cross-section which leads to generation of intertidal zones at this location (Figure 3-7b). This behavior after the closure is very similar to what Elias et al. (2003) indicated as the effect of the construction of the closure dam (Afsluitdijk).

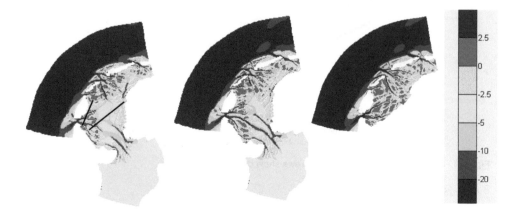

Figure 3 - 6 From left to right :Resulting bathymetry at the year 2000, at the year 4000 without the closure, and at the year 4000 with the closure dam included in year 2000

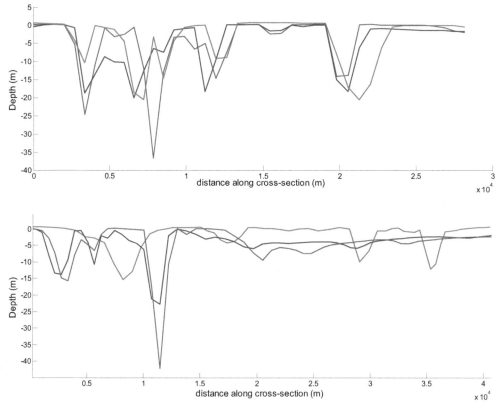

Figure 3 - 7 Resulting channel cross-sections at the year 2000(Blue), at the year 4000 without the closure(Red), and at the year 4000 with the closure dam included in year 2000(Green) - location of these cross-sections are shown in Figure 3 - 6 and the distance along cross-sections is measured from the west. (Top: close to the inlet, bottom: close to the closure dam)

3.6.2   Effect of closure on the tidal wave propagation

To identify the effect of the closure on the tidal wave propagation inside the basin we ran hydrodynamic model simulations at the morphological year of 2000 for both cases: with and without the closure. The tidal water levels at the time of high and low water at the Texel inlet together with the envelope of the tidal levels are shown in the Figure 3-8. The horizontal axis in this figure corresponds to the main channel of the Marsdiep basin prior to closure. The tidal wave propagating through the shallow Zuiderzee decreases and deforms by the bottom friction. Therefore the tidal amplitude at the southern boundary is considerably lower than the amplitude at the Texel inlet. Also Figure 3-8a shows that prior to the closure the tidal levels at the Texel inlet and at the southern boundary are out of phase. Due to the closure dam the length of the basin reduces considerably therefore the damping of the tidal wave also reduces. This less damped tidal wave together with the strong reflected wave from the closure dam forms a tidal wave with standing wave character and larger range. The change of the tidal range in the whole domain is shown in Figure 3-8c. As shown in the figure, the effect of the closure on the tidal range is more in the vicinity of the closure dam. The tidal range at this location is almost doubled after the closure.

3.6.3   Import / Export regime

At each closure time (1000, 2000, 3000 and 4000 years) we carried out two hydrodynamic/sediment transport model simulations and compared the discharge of water and sediment passing though the Texel inlet and the water level at the southern boundary of the inlet before and after the closure. Figure 3-9 shows the outcome of this comparison for closure after 1000 years and 3000 years of simulation. We can observe tidal asymmetry in the flood and considerable reduction in ebb-Sediment transport due to the closure. Therefore we can conclude that before the closure, the Texel inlet is ebb-dominant both for tidal-flow and tidal transport, thus, exporting sediment. After the closure Texel is still ebb-dominant for tidal flow but flood dominant for tidal transport, implying sediment transport into the Waddenzee basin. This result is in agreement with the short (1 month) simulations of flow and sediment transport at the time of closure, as reported by Elias et al. (2004). This change from being ebb-dominant to being flood dominant for tidal transport, changes the regime of the tidal basins in Waddenzee from a sediment exporting system to a sediment importing system.

Figure 3-10 shows the sediment exchange between North Sea and Waddenzee through the Texel and Vlie inlets in 4000 years morphological simulation without the closure together with the change of the import/export regime due to the closure at different times. The values are calculated based on the changes in the bathymetry of the basins, where a negative trend indicates export and a positive trend indicates sediment import. The result of the simulation shows that in the case of a Waddenzee without the closure dam, at the same time that the Marsdiep is ebb-dominant both for tidal flow and sediment transport, the whole system exports about 875 Mm^3 of sediment. The rate of this export is reducing in the course of time. In the case where the Zuiderzee is closed, regardless of the time of closure, the system starts importing sediment with the average rate of 0.5 Mm^3/Yr. over a period of 2000 simulation years without any sign of reaching "equilibrium". Clearly the amount of available sediment and morphology of initial bathymetry of the model has a significant effect on the rate of sediment exchange but the trends stay the same for the four scenarios.

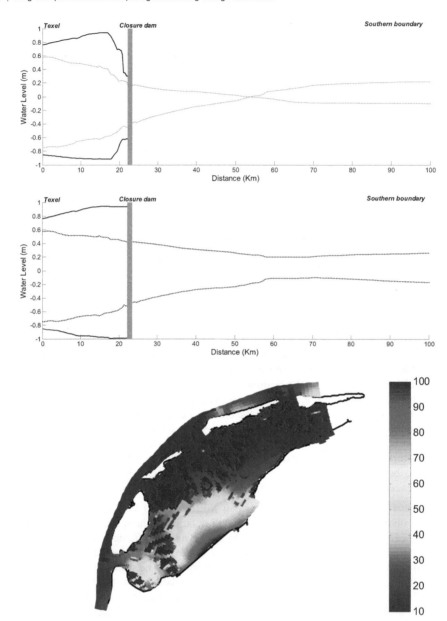

Figure 3 - 8 Simulated tidal water levels at the morphological year of 2000 prior and after closure. top: at the time of high and low tide at the Texel. middle : the envelop of tidal levels. bottom: The changes of the tidal range due to closure (%). (The dark red patches are the intertidal areas)

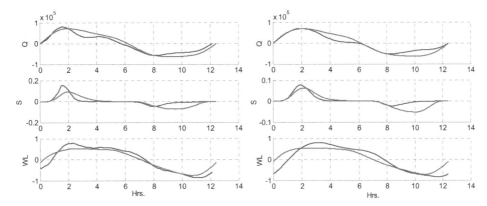

Figure 3 - 9 Discharge (Q), Sediment Transport (S) and water level (WL) in Texel inlet during a morphological tide before (red) and after the closure (blue) at year 1000 (left) and 3000 (right). Positive values of discharge and sediment transport are for the duration of flood

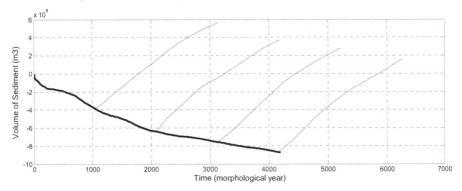

Figure 3 - 10 Sediment Exchange between North Sea and Waddenzee in 4000 years morphological simulation and the change of the import export regime due to the closure at different times, (+:import, - : export)

3.6.4 Intertidal flat characteristics

In literature, characteristics of the tidal basins such as relative intertidal flat area, intertidal flat height, intertidal flat volume, etc. are defined (references) and the evolution of the tidal basins can be seen through the changes in these characteristics. In this section, the changes in some of these characteristics during the 4000 year simulation and the effect of the closure on them are discussed. It should be noted that the domain which is considered to calculate these characteristics is the area covered by Marsdiep and Vlie basin after the closure (figure).

The intertidal flat is defined as the area between MLW and MHW; the existence of these areas in a tidal basin is what distinguishes a tidal basin from a submerged basin and make the tidal basins a unique type of coastal feature. The main characteristics of the intertidal flats are the area that they cover (area between MLW and MHW), the volume of sediment in these areas (volume of sediment between MLW and MHW), and the average height the intertidal flats calculated as :

$$h_f = \frac{V_f}{A_f}$$

In which,

Af	[m^2]	:Area of the intertidal flats
Vf	[m^3]	:Volume of intertidal flats
hf	[m]	:Height of intertidal flats

Figure 3-11 shows the expansion of the intertidal flats during the 4000 years simulation without the closure and the effect of closure on development of these areas. It is clear that initially, due to the initially flat bathymetry of the model, there are no intertidal flats present and that during the 4000 years of the simulation these areas expand. The rate of the expansion decreases with time which shows that the surface area of intertidal flats tends to reach an "equilibrium" value. At the time of the closure we can observe an initial jump in the value of the intertidal area and volume. This abrupt change is due to the definition of the intertidal flats and the changes in the tidal wave propagation. The intertidal flat is defined as the area between MLW and MHW and the closure causes an instantaneous change in the tidal range at different locations of the basin. Therefore some areas which were under the MLW, after the closure are inside the new tidal range, thus considered as intertidal flats. After this instantaneous change, intertidal flats expand at a higher rate than the rate of expansion before the closure. This new rate of expansion decreases with time. The rate of the expansion of the area of intertidal flats after the closure is dependent on the time of the closure. In the case where the closure happens after 1000 years of simulation, this rate is 3 times higher than the original rate in the simulation without the closure, while in the simulation with later closure times the average new rate of expansion is smaller. This is due to the fact that at year 1000 the intertidal flats are not developed enough and if the closure happens at this time the basin needs follow its natural expansion of intertidal flats and also cope with extra intertidal flats needed due to the closure. Eysink (1990) claims that one of the first parameters that tends to equilibrium on a relatively shorter timescale is the height of the flats, which is related to the tidal amplitude. In the simulation without the closure, the average height of the intertidal flats reaches its "equilibrium" state of about 0.6 m after 2000 years (Figure 3-10, lower panel). After the closure this value also increases with a substantial rate, this increase can be associated to the fact that in our simulations no wave generation takes place inside the basin. De Vriend et al. (1989) argue that the erosion of tidal flats due to these locally generated waves is in dynamic equilibrium with the sediment accumulation due to the tidal flood current. Therefore neglecting wave generation in the simulations leads to a higher sediment volume on the intertidal flats and consequently larger average intertidal flat height.

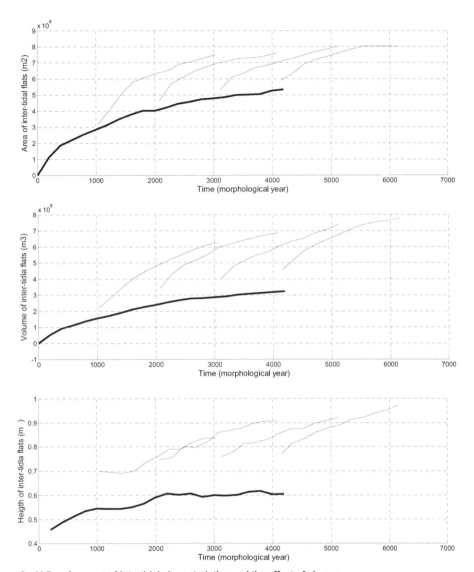

Figure 3 - 11 Development of Intertidal characteristics and the effect of closure

3.6.5   Texel ebb-tidal delta development

The other part of the tidal inlet system which is influenced by the construction of the closure dam is the ebb-tidal delta. The development of the ebb-tidal delta can be measured based on its volume (Walton and Adams 1976). The changes in the volume of the ebb-tidal delta of the Texel inlet during the different simulations with and without the closure are shown in Figure 3-12. It is clear that in the simulation without the closure, in which the basins continuously export sediment, the ebb-tidal delta gets larger and larger, initially (in the first 700 years) the expansion rate of the ebb-tidal delta volume is about 1 Mm³/year. Towards the end of the simulation this rate reduces to less than 0.08 Mm³/year. When the closure dam is constructed and the basins start importing sediment, the ebb-tidal delta loses sediment and regardless of the time of closure it reaches a new near equilibrium volume. This new equilibrium volume is dependent on the volume of the ebb-tidal delta at the time of closure.

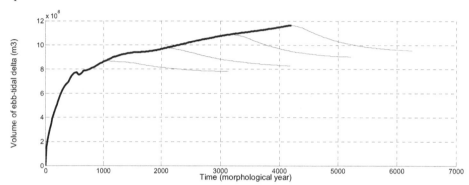

Figure 3 - 12 Development of the ebb-tidal delta and the effect of the closure

3.6.6   Boundary of the basins (Tidal divides)

Elias et al. (2003) have observed that the Marsdiep and Vlie basins, which used to be a single tidal basin with two inlets, are acting as almost independent basins after the closure of Zuiderzee. This means that due to the construction of the closure dam a tidal divide should develop between these two basins. In this section we analyze the result of the numerical simulations to show the same behavior in the model outcome. Although the physical boundary between the tidal basins is the tidal divide, the definition should be translated to flow field or bathymetry characteristics to find this divide in the results of the simulations. Following the definition in Chapter 2, we consider the line of minimum standard deviation of (depth averaged) velocities over a tidal cycle as the tidal divide. To study the change of the tidal divides, we followed the scenario in which the closure is applied after 2000 years of morphological evolution. Resulting bathymetry of numerical simulation is extracted:

a) At the morphological year of 2000 without the closure

b) At the morphological year of 4000 without the closure

c) At the morphological year of 2000 at the time of the closure

d) At the morphological year of 4000 including the closure dam at the year 2000

Then, in order to provide the corresponding flow field for all these conditions, using the extracted bathymetries, hydrodynamic simulations are carried out for 5 tidal cycles. Having the bathymetry and flow field in each case, using the above-mentioned definition it is possible to define the boundaries between tidal basins.

Figure 3-13 shows standard deviation of depth averaged velocities over a tidal cycle for each condition. In this figure it is clear that after 2000 simulation years the two basins are completely connected (Figure 3-13a). In the case without the closure the basins remain connected for the next 2000 years (Figure 3-13b). But at the year 2000 and exactly after the closure the tidal wave inside the basins changes and a wage boundary is established between the basins (Figure 3-13c), because of the instantaneous change in the flow field. This change starts the morphological changes and after the next 2000 years the morphology of the basins adapts to the new flow field and a distinguishable tidal divide is formed between the Marsdiep and Vlie basins (Figure 3-13d)

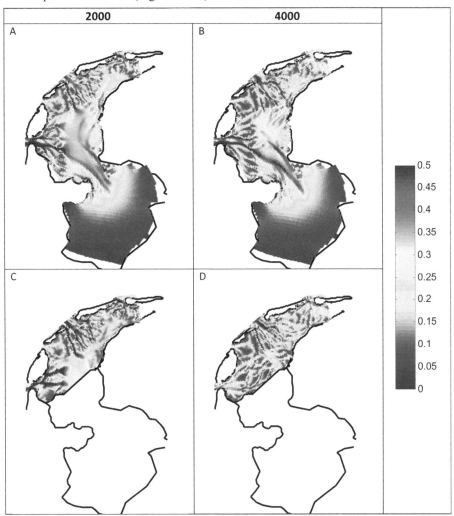

Figure 3 - 13 Standard deviation of depth averaged velocities over a tidal cycle A. At the morphological year of 2000 without the closure; B. At the morphological year of 4000 without the closure; C. At the morphological year of 2000 at the time of the closure; D. At the morphological year of 4000 including the closure dam at the year 2000.

3.7   Conclusion

In this chapter, we presented a long-term process-based modeling approach to study the effect of building a closure dam on the morphology of a tidal basin. The closure of the Zuiderzee is taken as a case study. The hydro- and morphodynamics for a total of five different scenarios were studied: a base scenario where no closure takes place, and four scenarios corresponding to different times of closure.

The model is capable to reproduce the change in tidal transport regime and the ensuing changes in morphological characteristics of the tidal basins. Furthermore, the model shows that regardless the time of closure, the surface area of the intertidal zones and the volume of the ebb-tidal delta tend to an equilibrium value. Also we have shown that the new " hydrodynamic tidal divide" changes instantaneously and then the "morphological tidal divide will follow".

The results prove that long-term process-based modeling is capable to at least qualitatively assess the long-term impacts of large scale human intervention in a coastal system.

CHAPTER 4

Mixing the soil

Effect of sediment see bed composition in long-term morphological modeling of tidal basins

4.1   Introduction

In recent years process-based models have been used more and more to simulate the morphodynamic behavior of tidal basins at various temporal scales (Wang et al. 1995, Hibma et al 2003, Marciano et al 2005, Ganju et al 2009). These studies show that process-based morphological models, describing in detail the flow field and the resulting sediment transports and bottom changes, perform well in simulating the complex morphodynamics of tidal basins. Later Van der Wegen and Roelvink (2008) and Van der Wegen et al (2008) showed that by using the 'online approach' in morphological modeling it is feasible to use a process-based model for longer-term (~centuries and ~millennia) morphological simulations. This approach is used also in chapter 2 to simulate the long-term evolution of tidal basins. Morphological developments of tidal basins and estuaries in medium or long time scales such as relocation of shoals and channels in response to human interventions (e.g. dredging) or natural disturbances (e.g. sea level rise) are the main input for the decision makers in the fields of coastal zone management. Therefore, considerable efforts are currently being undertaken to hindcast and/or forecast these developments. In addition, ecological and economical values and functions of tidal basins are closely linked to their morphodynamic developments.

In most cases this type of morphological simulations needs considerable computational time (in the order of weeks). Therefore the schematization of model inputs together with acceleration techniques plays a significant role in making these simulations feasible. In recent studies some successful schematization strategies in hydrodynamic boundary forcing are introduced, such as the concept of the morphological tide (Latteux, 1995 later modified by Lesser, 2009) and reducing the number of wave conditions (Lesser, 2009). Also acceleration techniques such as applying a morphological factor (Roelvink 2006) are widely used.

However, the resulting bathymetry of the hindcasting simulations usually does not match the measured data perfectly. In the case of tidal basins these simulations generally result in too deep and narrow channels and too high shoals in tidal basins. This type of shortcomings is due to the sediment characteristics which are introduced in the model. Usually in these simulations one type of sediment is used and the sediment size is represented by a single D50 grain size. Therefore despite their importance in morphological changes, some mechanisms such as armoring of the channel bed due to erosion of finer materials and sediment sorting are systematically neglected. If sediment mixtures are needed to be applied in the model, sediment characteristics and their initial distribution over the model domain are required. However, they are at best measured at yearly or even decadal intervals. This especially holds for data on bed composition and sediment properties, if measured at all. In order to find a solution for this problem, previously we have introduced the concept of 'morphodynamic spin-up' as an analogue to the 'hydrodynamic spin-up' which refers to the time period that is required for the hydrodynamics within a model domain to adapt to the boundary condition forcing. Similarly, the morphodynamic spin-up describes a time period in which the morphology adjusts itself to forcing conditions of the morphodynamic model (Van der Wegen et al 2011). This morphodynamic adaptation can be in terms of bed level change but also in terms of bed composition. The time scale of the 'morphodynamic spin-up' is considerably larger (up to five orders of magnitude) than the time scale of the 'hydrodynamic spin-up'. To avoid the unrealistic morphological changes due to 'hydrodynamic spin-up' time, usually the bed level updates are frozen during the 'hydrodynamic spin-up'. Also we have suggested a pre-run called Bed Composition Generation (BCG) run as 'morphodynamic spin-up' to redistribute different types of sediment in the model domain without changing the morphology. The actual morphodynamic calculations are then carried out with the bed composition resulting from the BCG run (Van der Wegen et al, 2011).

4.2   Aim of the study and methodology

The main goal of this study is to show the importance of including sediment mixtures in long-term morphological models, and to develop procedures to compensate the lack of bed composition measurement data. To fulfill the goal of this study, a numerical, process-based model (Delft 3D, Lesser et al (2004)) is used that allows for the definition of different types of sediment (classes of sediment). In the first part of the paper the importance of using different mixtures of sediment in long-term simulations of tidal basins is illustrated. Then by using the outcome of these long-term simulations, different procedures are developed to derive an initial sediment distribution over the model domain. To compare these procedures, we derived initial sediment distributions for the area of interest based on each procedure, and carried out hindcast simulations for 75 years (1930 to 2005). The results of these simulations are compared with each other as well as with the conventional methods of using a single sediment type.

4.3   Area of interest and available data

The area which is chosen as the case study in this research is the Dutch Waddenzee. The Dutch Waddenzee is the western part of the Waddenzee in the south east side of the North Sea along the Netherlands coastlines. This part of the Waddenzee consists of four tidal inlets connecting the North Sea to the basins, behind the barrier islands. (Figure 4-1). The Dutch Waddenzee was subjected to a major human intervention in 1932. In this year the south part of the basins (Zuiderzee) was closed by a 32 km dike known as the Afsluitdijk. Due to this closure, the area of the tidal basin was reduced by around 3300 km2 and the length of the basin by 100 km. This intervention triggered substantial morphological changes in this part of the Waddenzee. Since the Waddenzee is one of the best monitored tidal basins in the world and there are lots of historical data available, it is a very interesting area to be used as a case study for hindcasting simulations.

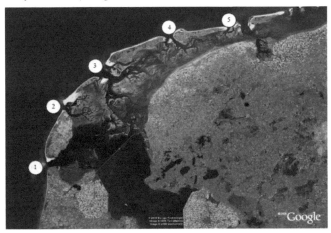

Figure 4 - 1 Satellite image of Dutch Waddenzee: 1- Texel-Marsdiep 2- Eierlandse Gat 3- Vlie 4- Amelander Zeegat 5- Frisian inlet.

Due to the extensive maritime transport in the Netherlands since the 16th century, there are some historical depth measurements available from that time. After the construction of the Afsluitdijk the depth measurements were carried out more often in this area and since 1987, Rijkswaterstaat (Directorate-General of Public Works and Water Management of The Netherlands) has frequently measured the bed level in the Waddenzee. The ebb-tidal deltas are measured every 3 years, while the basins are measured every 6 years. These data are stored in a 20 x 20 m resolution database called 'Vaklodingen'. Recently all the available data since 1926 has been collected, processed and made accessible via the OPENEARTH project (www.openearth.nl). Using this set of data it is possible to construct the historical bathymetries for different parts of the Dutch Waddenzee. Figure 4-2 shows the bathymetry of this area in the year 1926 (date of the closure of the basins).

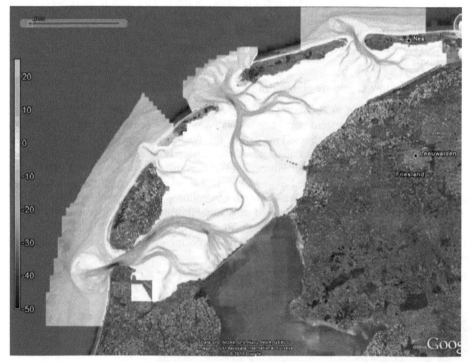

Figure 4 - 2 Bathymetry of the Dutch Waddenzee in 1926 (projected on the google earth image using OpenEarth tools)

Also in the period between 1989 and 1997 a total of more than 7000 soil samples were taken from the bottom of the Waddenzee. The soil samples were taken from the upper 10 cm of soil. Initially one sample was taken per square kilometer, but in areas with a rapidly changing morphology a finer grid is used and one sample is taken per half square kilometer. A Malvern Laser Particle Sizer is used to determine the grading of these samples and for each sample a grading curve was produced. All of these data are available in the Sediment Atlas Waddenzee and also via the OPENEARTH project. Figure 4-3 shows the map of the median grain size (D50) distribution in the Dutch Waddenzee. As can be seen in Figure 4-3, a wide range of sediment sizes are present in the area with a maximum of more than 0.5 mm, as expected, in the channels and the a minimum of 0.03 mm on the shoal areas close to the land boundaries of the basins.

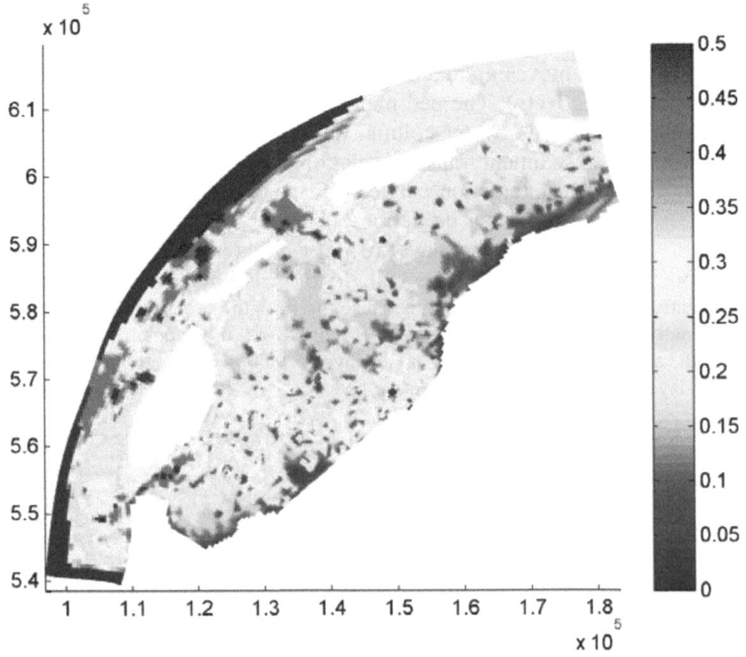

Figure 4 - 3 Map of the median grain size (D50) distribution in the Dutch Wadden in mm

4.4   Model Description

The model used in this study is the depth averaged version of Delft3D. This model is extensively described by Lesser et. al. (2004) and Van der Wegen and Roelvink (2008). The model uses a finite difference-scheme, which solves the momentum and continuity equations on a curvilinear grid with a robust drying and flooding scheme the velocity field obtained by solving the equation of continuity and the momentum equations is used to calculate the sediment transport field. Every time step, as a consequence of the divergence of the sediment transport field, the bed level is updated. In this approach the flow, sediment transport and bed-level updating run with the same (small) time steps. Since the morphologic changes are calculated simultaneously with the other modules, coupling errors are minimized. But, as described in Lesser et al. (2004), because this approach does not consider the difference between the flow and morphological time step, a 'morphological factor' should be applied to increase the rate of depth changes by a constant factor in each hydrodynamic time step (Lesser et al., 2004, Roelvink, 2006). A bed layer model based on the concept of the active layer (Hirano, 1971) is included in the model, therefore different sediment classes can be defined and different transport formulation can be assigned to each sediment fraction. More details about the bed layer model are described in the following section.

4.4.1   Bed layer model

Blom (2008) describes four different sediment continuity models for non-uniform sediment in comparison with a laboratory flume test. Among them she also discussed the concept of an active layer introduced for the first time by Hirano (1971). The same concept is applied in the current study with a slight adaptation. The grid cells in the bed layer model use the same grid cells as the hydrodynamic model in the horizontal direction, dividing the bathymetry into the same number of cells. A thickness of sediment is assigned to each one of these cells representing the initial available sediment in that cell. In each cell there is an active layer with a predefined thickness which moves up and down with the bed level of each cell (in case of deposition and erosion respectively). The bed material, which can be composed of different sediment classes, interacts with the water column through this active layer. Each sediment fraction follows the specified sediment transport relation assigned to that fraction. The model does not consider any physical interaction between different classes. Underneath the active layer, some numerical under-layers can be defined to allow for a varying bed composition with depth. Figure 4-4 shows how the bed layer model works with different sediment classes in the case of deposition or erosion in one cell. In the case of erosion (Figure 4-4a) the composition of the active layer becomes coarser, because the fines are eroded and the coarse material is left behind in the active layer. In case of deposition (Figure 4-4b), the deposited fine material make the composition of the active layer finer. A new under-layer develops with a composition that is finer than the other under-layers, but coarser than the active layer.

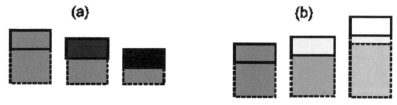

Figure 4 - 4 (a) Erosion process of bathymetric cell; (b) accretion process of bathymetric cell; Solid line represents active layer. Dotted line represents sediment layer. Darker color indicates coarser sediments and less shading indicates finer sediments.

4.5   Long-Term model set up

In order to show the effect of different sediment mixtures on the simulation of the evolution of tidal basins, a long-term model (~500 yrs) was set up. The model used for these simulations is the same model that has been used in Chapter 2. The grid extends from the south of Texel inlet to the tidal divide between Amelander Zeegat and Friesche Zeegat from the other side (Figure 4-5). The average size of grid cells inside the basins is 350 m, with the seaward boundary 46 km offshore. The computational grid does not include the dry part of the barrier islands (areas with an elevation above high water).

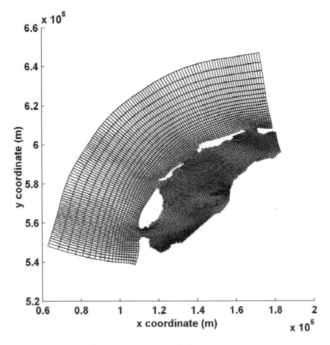

Figure 4 - 5 Computational grid for long-term simulations

In this model we have considered only the tidal forcing by adopting the concept of morphological tide (Latteux, 1995). Therefore the spring neap cycles are ignored and only representative M2, M4, M6 components of the tide are applied on the seaward boundary of the model. The lateral boundaries are Neumann boundaries, that prescribe M2, M4, M6 alongshore water level gradients (Roelvink and Walstra, 2004).

A simplified bathymetry is used as the initial bathymetry of this model. To generate this bathymetry it is assumed that there are no channels and shoal areas inside the tidal basins and the whole area is completely flat with a constant depth of 4.54 m (average depth of the basins). Applying this initial bathymetry allows the model to show the evolution of the basins without being constrained by the initial position of channels or shoals. Therefore the model strives to find a stable solution according to the cycle of hydrodynamic forces, sediment transport and redistribution, and bed level changes. This model is used to carry out simulations with different mixtures of sediment classes.

4.5.1 Selection of sediment classes and bed layer model setup

The long-term simulations include 6 sediment classes. These classes are based on the standard sieve test and are represented within the model by the minimum and maximum grain size of each fraction. It is assumed that the sediment size distribution of each fraction has a log-uniform distribution (Table 4-1). No mud fraction is included in the sediment mixtures.

Table 4 - 1 Sediment grain size in each sediment fraction.

Fraction	Minimum size (mm)	Maximum size (mm)
1	0.075	0.150
2	0.150	0.300
3	0.300	0.425
4	0.425	0.600
5	0.600	1.180
6	1.180	2.360

These sediment classes are mixed with different percentages to provide a sediment mixture with the D50 similar to the overall measured D50 (0.250 mm). Figure 4-6 shows model input grading curves together with the range of the measured data and the overall grading curve of all the samples in the area.

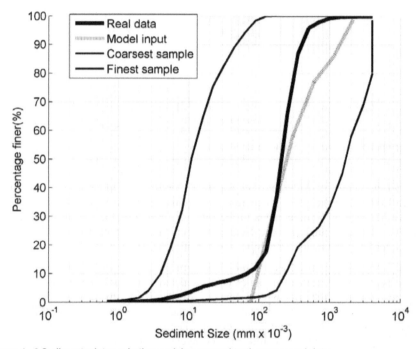

Figure 4 - 6 Sediment mixtures in the model compared to the measured data

A simulation is carried out with these sediment mixtures, which initially have a uniform spatial distribution over the domain. Therefore each bed cell consists of all 6 sediment classes with different portions of the bed cell volume. In each cell the sediment is available down to a depth of 65 m below mean sea level. For the sake of comparison, one simulation is also carried out with only one fraction of sediment with the same D50 as the sediment mixtures (0.250 mm).

Blom (2008) suggested that the best results of Hirano (1971) model can be obtained when the thickness of the active layer is similar to the dune height. However when a large morphological factor is applied in the simulation, to ensure the numerical stability of the model other consideration should be taken into account. A primary sensitivity analysis shows that the available sediment in the active layer should not be entirely replaced in one time step, therefore the thickness of the active layer should be more than the typical change in bed level in one time step. Since a relatively large morphological factor (300) is used in these simulations the active layer thickness of 1.5 m is chosen and 5 underlayers with a maximum thickness of 2.0 m are defined.

4.5.2   The effect of sediment classes on long-term morphodynamic simulations

Morphological evolution

Starting with the schematized, flat bathymetry inside the basin, the first 100 years of modeling shows mainly the development of ebb-tidal deltas in front of the inlets and the main channels in the inlets. Later, after almost 400 years, the main channel/shoal patterns inside the basins have almost fully evolved; however, the dynamic behavior of channels and shoals is obvious. This behavior is observed in all the simulations although the channels are shallower and wider in the simulations with a higher percentage of coarse sand compared to the simulation with only one fraction of sediment with the same D50. Figure 4-7 shows the initial condition and resulting bathymetries for two simulations for the Marsdiep tidal basin. This suggests that we should be able to see the effect of adding coarse sand in the characteristics of tidal basins such as the hypsometry.

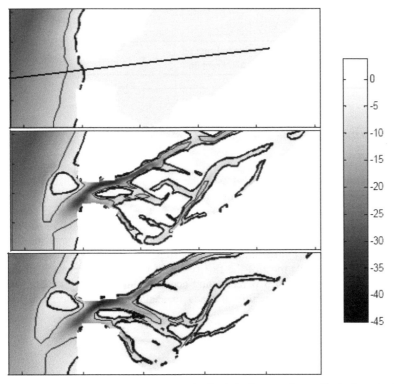

Figure 4 - 7 Simulated Evolution of Marsdiep Basin bathymetry from a flat bathymetry inside the basin, after 500 years of morphological modeling. (top : initial condition, middle : model with one sediment fraction, bottom : model with multiple sediment fraction)

Comparing the hypsometry

In order to compare the hypsometry for different simulations, we chose the Marsdiep tidal basin. The hypsometry curve of Marsdiep basin at the end of different simulations is shown in Figure 4-8. As it is shown, by adding more coarse sand to the model the tidal basin will have shallower channels and less high shoals.

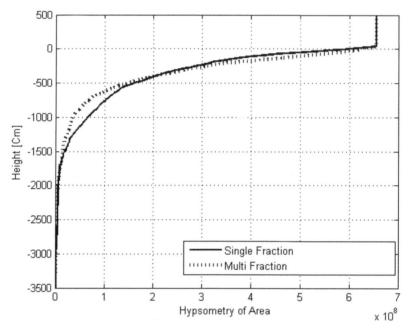

Figure 4 - 8 Hypsometry of Marsdiep Basin for two simulations

Sediment Distribution

In the simulations with a mixture of sediment classes, the sediments are also redistributed in the domain, so that the sediment composition in each bed cell changes. This change for the finest sand fraction in an arbitrary longitudinal cross-section of the Marsdiep basin is shown in Figure 4-9. It is clear that in the absence of waves, the finer sediments tend to move to the ebb-tidal delta and also the shoal areas inside the basin (increasing the percentage of the fine sediment in the sediment mixture from the initial value of 10 percent all over the domain up to 40 percent), while in the channels most of the fine sediment is eroded (percentage of the fine sediment is decreased to less than 5 percent) and coarse sediment remains there and makes the channel bed less erodible. In turn it leads to shallower and wider channels compared to single fraction runs.

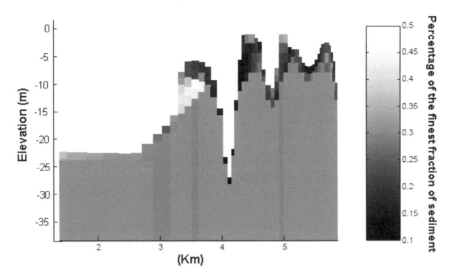

Figure 4 - 9 Percentage of the finest sand fraction in the sediment mixture after 500 years of morphological changes.(The location of the cross-section is shown in Figure 4 -7)

This redistribution of sediment classes can also be seen by the change in the average D50 in each grid cell. Figure 4-10 shows the resulting spatial distribution of the D50 in the Marsdiep tidal basin after 500 years of morphological simulation.

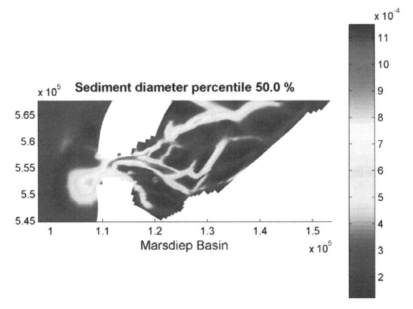

Figure 4 - 10 Spatial distribution of D50 in the Marsdiep tidal basin after 500 years of Morphological simulation. (in mm)

During these long-term simulations the model aims to reach a stable state. The sediment redistribution has a significant effect on the solution. These results show that the presence of different sediment sizes in a morphological model is important to obtain an acceptable result in (long-term) morphological simulations. On the other hand, one could imagine that the initially uniform sediment distribution over the domain has an unrealistic effect on the observed morphological changes. The next question is therefore how a more realistic initial sediment distribution over the model domain can be derived in the absence of adequate data .

4.6   Realistic hindcast with sediment classes

The previous section described the possible effect of different sediment classes on the morphodynamic evolution of the Waddenzee, albeit with a highly schematized, initially flat, bathymetry. The following section focuses on a more realistic hindcast of the morphodynamic developments from 1930 to 2005. Special emphasis is put on a methodology to derive a useful initial sediment distribution over the model domain based on a realistic, measured bathymetry. This methodology makes use of the runs described in the previous section.

The current research applies a new methodology to produce the initial conditions of the 1930 sediment distribution. Main assumption is that the resulting sediment distributions of the long-term simulations described in the previous section are in agreement with the measured data. To reach this similarity different sediment mixtures have been tested and a sediment mixture with 8 different classes has been introduced to the model. This mixture is coarser than the overall sediment mixture in the Waddenzee. However, during the 500 years of morphological developments, some of the coarser sediments are buried under layers of fine sediment. After 500 years the sediment grading curve of the sediment in the top layer is similar to the overall sediment grading curve of the measured data. This comparison is shown in Figure 4-11.

In order to use this long-term model, not only should the sediment grading resemble the measured data, but also the sediment distribution should be similar. To check the changes in sediment distribution, the standard deviation of the median grain size in the top layer is calculated during the 500 years of morphological development (Figure 4-12). The standard deviation is zero at the beginning, since every cell in the model has the same D50 (initial uniform spatial distribution of sediment classes). Due to the erosion, deposition and sediment redistribution, in the first 100 years the standard deviation increases rapidly, after 100 years it increases with a lower rate and finally after 400 years it is stabilizing. This suggests that, although the morphological evolution of basins does not stop after 500 years, the sediment classes are redistributed in such a way that the morphological changes do not have a significant effect on the overall sediment distribution. These results can be used as a proper basis to generate the initial sediment distribution for the hindcasting model.

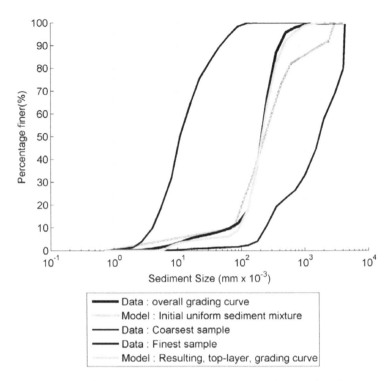

Figure 4 - 11 Range of measured sediment grading curves with the total grading curve, initial uniform sediment mixture and the resulting, top-layer, grading curve.

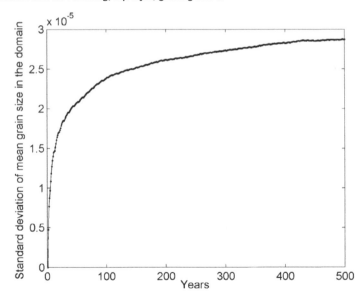

Figure 4 - 12 Development of standard deviation of mean grain size in the domain in 500 years of morphological modeling

4.6.1 Preparing the initial sediment distribution for the hindcasting model

The 500 year model results allow for characteristic relations between hydrodynamic parameters and grain size to be investigated. One of the best parameters that can be related to the grain size at each point is the bed shear stress. We have chosen the variance of depth averaged flow velocity over one tidal cycle as an indicator of the bed shear stress. A hydrodynamic simulation is carried out for the resulting bathymetry of the long-term simulation and the variance of depth average flow velocity is calculated for each point in the model. Figure 4-13 shows that the sediment median grain size is in a fair relation with the variance of velocity. An exponential curve with a maximum is fitted to the data. The maximum value of D50 in this relation is limited to the maximum D50 measured in the field (1.4 mm).

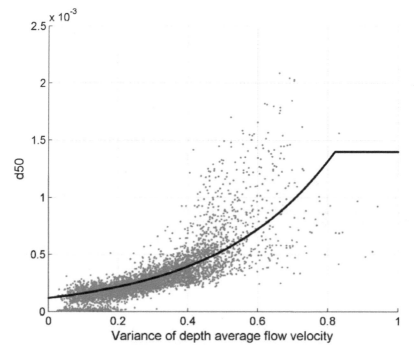

Figure 4 - 13 Relation between sediment size in each point and the variance of depth average flow velocity at the same point

Using this relation it is possible to produce an initial sediment distribution for the hindcasting model of the Waddenzee.

4.6.2 Hindcasting Model setup

The domain and forcing of the model which is used for the hindcasting simulation is the same as the model for the long-term simulations. Since more detailed morphological features are important in the hindcast simulation, a finer grid has been used for this model with an average grid cell of 100 m. The initial bathymetry for this simulation is the bathymetry of 1930 and the morphological simulation is of a 75 year period. To decrease the computational time a morphological factor of 75 is chosen, such that the hydrodynamic simulation time is one year. The model applies the bed layer model. Four different simulations have been carried out.

Simulation I

To produce the initial sediment distribution for this simulation, first a hydrodynamic simulation has been carried out for the bathymetry of 1930. For each point in the model, the variance of depth averaged velocity over one tidal cycle is calculated. Using the relation shown in Figure 4-13 the median grain size is determined for all the cells over the model domain. The map of spatial distribution of the D50 determined by this routine is shown in Figure 4-14.

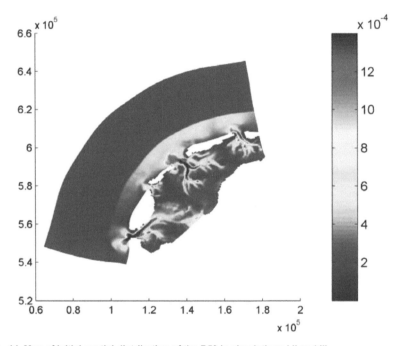

Figure 4 - 14 Map of initial spatial distribution of the D50 in simulations I,II and III

The spatial sediment distribution is considered to be constant over time. This means that in every cell of the bed layer model there is only one fraction of sediment with a constant grain size over time; clearly the cells in the channels contain coarser materials than the ones on the shoals. The shortcoming of this method is that when a sediment grain moves from one cell to another one, for example from a shoal to a channel, the size of the sediment will change automatically. If a channel with coarse sediment migrates to a former shoal area with fine sediment, the sediment size in the channel will become finer. This error will be more pronounced in case of large morphological changes.

Simulation II

The procedure which has been used in simulation II is exactly the same as the simulation I, but in order to reduce the effect of the error induced by large morphological changes, the spatial distribution of D50 in the model is updated every 5 years. This means that, at the end of each 5 years simulation intervals (1935, 1940, ...), a hydrodynamic simulation is carried out over the resulting model bathymetry for that specific year. Then the variance of depth averaged velocity over one tidal cycle is calculated in every grid cell and the median grain size is determined for all the cells in that year. The simulation then continues using the updated spatial distribution of the median grain size.

Simulation III

In both simulations I & II each bed layer cell contains only one fraction of sediment, so in principle there is no mixture of different sediment classes and redistribution of the sediment does not occur in the model. In Simulation III we use all the 8 sediment classes which have been used in the long-term model simulation in the model. Thus, each cell contains a mixture of the 8 sediment classes under the condition that the overall D50 of the cell follows the map presented in Figure 4-14. This leads to a more realistic model which allows the sediment to redistribute during the morphological simulation. For this purpose, the result of the long-term simulation is used to check the percentage of different sediment classes in each bed layer cell as a function of its D50 at the end of 500 years of simulation. Clearly larger D50 means higher percentage of coarser sediment classes. The graphs in Figure 4-15 show the percentage of each sediment fraction as a function of D50 for all of the bed layer cells in the model.

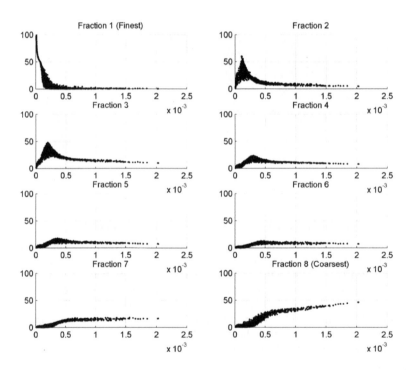

Figure 4 - 15 Percentage of each sediment fraction in each numerical cell as a function of D50 of that cell

As can be seen in the Figure 4-15, with the exception of finest fraction, there are two distinct parts in the data. To fit a curve to these data one curve is fitted to each part and then by the use of a shape function (Equation 1) both curves are merged into one relation. An example of this method for fraction 5 is shown in Figure 4-16.

$$f(x) = 1 - e^{-\left(\frac{x}{X}\right)^n}$$

(1)

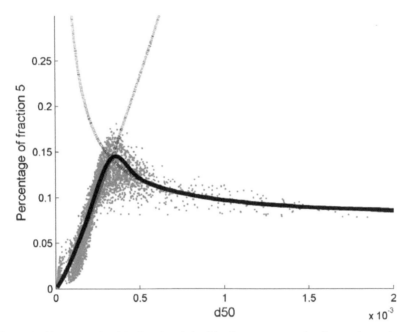

Figure 4 - 16 An example of the function derived for the percentage of sediment classes in numerical cells based on the D50 of the cells

By using these relations for each fraction, the D50 (Figure 4-16) in each cell is converted to a sediment mixture of all 8 sediment classes. These percentages of sediment classes in each cell are introduced in the model as the initial condition of the third hindcasting simulation.

Simulation IV

For the fourth simulation first a Bed Composition Generation (BCG) run (Van der Wegen et al, 2011) was carried out. As described previously, the idea is that, given a set of sediment classes distributed uniformly in the model, a special BCG run generates a stable distribution of sediments that fits the hydrodynamic conditions on a fixed bathymetry. In our case the BCG run was carried out for 500 years, with the same 8 sediment classes as the long-term simulations on the bathymetry of 1930. The result of the BCG simulation is a mixture of sediment classes in each bed layer cell. The outcome of this BCG run presented as the D50 in each cell is shown in Figure 17.

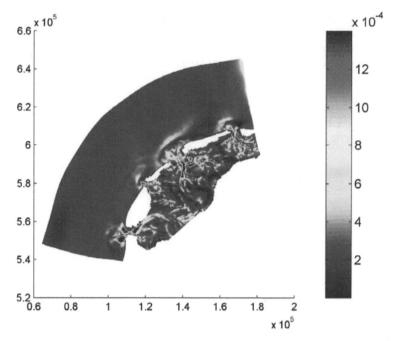

Figure 4 - 17 Map of initial spatial distribution of the D50 in simulations IV

4.6.3 Bench mark simulations

Two bench mark simulations are also carried out to show the effect of considering different sediment classes in this hindcast simulation. One simulation (referred to as B0) applies only one fraction of sediment with a D50 of 0.25 mm (the median grain size in the area)   and the second simulation (referred to as B1) includes all 8 classes of sediment with a uniform spatial distribution.

4.7   Results and Discussions

In this part the result of the different simulations are compared with each other and to measurements. First the skill score of the simulations is calculated to show that the skill of the model improves by using different sediment classes and an initial sediment distribution. Later the simulated morphological changes are discussed based on the erosion-deposition patterns in the Marsdiep basin and channel cross-sections. Finally the effect of different sediment classes on better representing other process involved in the model is investigated.

4.7.1   Brier Skill Score (BSS)

The BSS is commonly used in meteorology models and recently has been used in the evaluation of the coastal morphodynamics models by Brady and Sutherland (2001), Sutherland et al. (2001), Van Rijn et al. (2002), Van Rijn et al. (2003) and Sutherland and Soulsby (2003). Sutherland et al. (2004) introduced the following relation to calculate the BSS for morphological models :

$$BSS = 1 - \frac{< (Y - X)^2 >}{< (B - X)^2 >}$$

(2)

In which Y is a prediction, X is an observation, and B is a baseline (either prediction against which the model skill will be evaluated or initial bathymetry). In this relation , $< >$, means taking an arithmetic average. In other words BSS provides an objective quantification of the skill of a model, where skill is defined to be the accuracy of a model prediction relative to a baseline prediction. The skill score calculated in this manner will have the maximum of 1.0 which means the perfect match between prediction and observation. The lower limit of the skill score is not bounded to any number and The BSS can become negative if the model result is less accurate than the baseline model. The value of zero for BSS indicates that the model prediction is the same as the baseline. In the case that initial bathymetry is considered as the baseline, van Rijn et al.(2003) proposed a qualification of BSS, as repeated in Table 4-2.

Table 4 - 2 Classification of the morphological models based on the BSS

Qualification	BSS
Excellent	0.8 – 1.0
Good	0.6 – 0.8
Reasonable	0.3 – 0.6
Poor	0 – 0.3
Bad	<0

The BSS is calculated for the resulting bathymetry after 75 years for all six simulations. In this calculation the bathymetry of 1930 (initial bathymetry of the simulations) is considered as the baseline (B) and the measured bathymetry of 2005 projected on the computational grid as the observation (X). Since there is no wave forcing in the model, the results of the simulations for the ebb-tidal deltas and the outer coastline are probably not adequate and these areas are excluded from the calculation of the BSS. The tidal basins are also divided according to the tidal divides and the BSS is calculated for each basin separately. The BSS of the different simulations for each basin is summarized in Table 4- 3.

Table 4 - 3 Brier skill score of each simulation for different tidal basins

Simulations	Basins		
	Marsdiep	Eierlandzegat	Vlie
B0	-2.81	-0.37	-0.92
B1	-1.68	-0.24	-0.58
I	0.08	-0.06	-0.04
II	0.06	-0.08	-0.05
III	0.08	0.08	0.04
IV	-0.04	0.11	-0.01

Table 4-3 shows the improvement of the model when an initial sediment distribution is considered in the model. Simulations B0 and B1 have a negative skill meaning that the initial bathymetry better resembles the developed bathymetry than the modeled bathymetry. When an initial spatial distribution is introduced in the model (Simulation I-IV) the BSS highly improves and becomes positive for almost all basins. However the BSS value shows that the model results are still poor. Additionally this table shows that regardless of how the initial spatial sediment distribution is provided the skill scores of the simulations are in the same range.

To look into the differences made by using the sediment classes and to explore the reasons for the poor BSS we will discuss the morphological changes in one of the basins (Marsdiep) for the best (III) and the worst (B0) simulations in the next paragraphs.

Figure 4-18, shows the amount of sedimentation and erosion in the Marsdiep basin from 1930 to 2005 in measured data and simulations. This figure shows that the most pronounced difference between the two simulations is in the depth and width of the channels. These differences in the channel shape also become clear by comparing the results for a channel cross-section (Figure 4-19). In bench mark simulations the channels are getting unrealistically deep whereas the sediments from the channels are deposited on the channel banks so that the channels become narrower. The rate of this process is much lower in the simulation with sediment distribution and therefore leads to more realistic prediction of the position and shape of the channels in this simulation and consequently a higher skill score.

However in both simulations the model did not predict the deposition at the end of the old channels, close to the closure dam and south part of the basin. This may be due to two main reasons. First, there is no mud in the model and therefore the suspended sediment transport, which can be responsible for this deposition is neglected in the model; the fact that the deposition in these areas has a high mud content can explain this discrepancy. The second reason may be the absence of waves in the basins. De Vriend et al (1989) suggest that these locally generated waves can have significant effect on the flat areas. Furthermore, due to the outflow of fresh water from the IJssel Lake, vertical recirculation may lead to more fine sediment import and deposition.

The migration of the main channel in the basin to the north is underestimated in the model; this may be due to 3D flow effects such as spiral flow, which were neglected in these simulations.

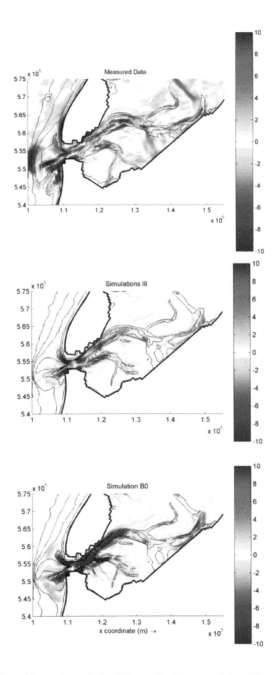

Figure 4 - 18 Erosion deposition pattern of simulations after 75 years of simulation compared with the measurements between 1930 and 2005

4.7.2   Cross Sections

In the last section we claimed that the simulation with sediment classes and initial spatial sediment distribution predicts the shape of the channels better than a simulation with a single uniform sediment fraction. In this section we choose one cross-section of Marsdiep basin and compare the model results with the measured data of 1930 and 2005. Figure 4-19 clearly shows that the simulation with a single sediment fraction (B0) predicts a much deeper and narrower channel than the other simulations.

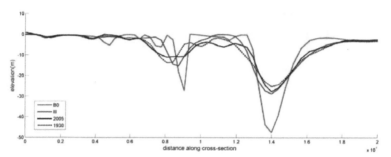

Figure 4 - 19 Resulting cross section after 75 years of morphological simulation for the models with (Blue) and without (Red) initial spatial sediment distribution compared with the measured cross-section in 1930 and 2005

Figure 4-20 shows the result of all simulations with spatial sediment distribution. It can be concluded from this figure that the procedure for producing the initial condition does not have a significant effect on the resulting depth and width of channels.

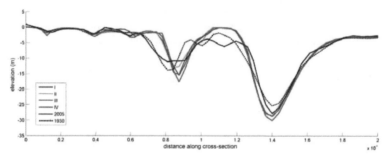

Figure 4 - 20 Resulting cross section after 75 years of morphological simulation for the models with different initial spatial sediment distribution simulations compared with the measured cross-section in 1930 and 2005

Figure 4-21a shows the mean sediment size (D50) for simulations B1 and simulation III. Both simulations apply the same sediment classes, however in the first one the sediment mixture is uniform on the whole model domain and in the second simulation an initial distribution is introduced in the model. Clearly the initial value for this parameter is constant (0.25 mm) in simulation B1, after 75 years of morphological simulation, distribution of D50 is following the initial distribution introduced to simulation III. But this adaptation is happening together with the morphological changes and it leads to more than 10 m of unrealistic deepening of the main channels of the basin (Figure 4-21b). Therefore for long-term morphological simulations it is important to cope with the concept of 'morphological spin up' by introducing an intelligent guess for the initial distribution of sediment grain size.

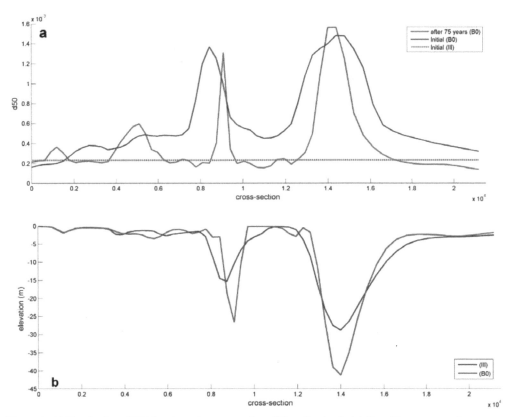

Figure 4 - 21 Distribution of D50 in a lateral cross-section of Marsdiep basin (a). Resulting cross section after 75 years of morphological simulation (b)

4.7.3 Effect of using sediment mixtures on the impact of the bed slope on the adjustment of the sediment transport rate

Another interesting parameter to investigate is the impact of the bed slope on the adjustment of the sediment transport rate. This adjustment is referred to as the 'bed slope effect'. In order to account for the effect of a transverse bed slope on the rate and direction of bed load sediment transport usually the Ikeda (1982) suggestion is used in the models:

$$S_n = \left|S'\right| \alpha_{bn} \frac{u_{b,cr}}{\overrightarrow{\left|u_b\right|}} \frac{\partial z_b}{\partial n}$$

In this relation the α_{bn} is the tuning parameter with the suggested value of 1.5; however in long-term morphological simulations, to compensate for neglected processes and obtain reasonable results, this value is increased to 10 (Van der Wegen 2011). To check the effect of using a mixture of sediment on the value of this parameter, simulation III has been carried out with two different values for α_{bn}: 1.5 and 10. The resulting bed levels for the same cross-section as in Figure 4-20 are shown in Figure 4-22. This figure shows that although the value of α_{bn} affects the morphological development, this effect is not considerable when sediment sorting is taken into account.

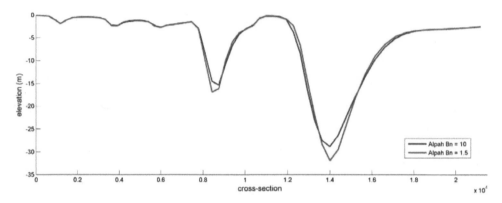

Figure 4 - 22 Resulting cross section after 75 years of morphological simulation for different α_{bn}

In this research there are quite a number of processes which are neglected, such as the effect of the wind generated waves inside the basin and the effect of biological processes on the morphology of the tidal basins. Including these processes may increase the skill of the models and needs to be investigated.

4.8   Conclusions

This research clearly shows that when we are dealing with long-term morphological simulations the sediment composition plays a significant role. Introducing a logical initial sediment size distribution improves the performance of the model significantly. This initial sediment size distribution is estimated by adapting the sediment distribution to the hydrodynamic condition at the beginning of the simulation and can be produced by relating the sediment size to the bed shear stress as well as carrying out pre-simulation runs to account for the 'morphological spin-up' time. Also it is shown that application of this method can reduce the need to artificially enhance coefficients such as the bed slope effect.

CHAPTER 5

Switching channels

The effect of waves and tides on the morphology of tidal inlet systems

5.1 Introduction

Barrier islands and tidal inlets are found along about 15% of the world's coastlines (Davis and Fitzgerald, 2004). Tidal inlets separate successive barrier islands and provide a connection between the tidal (back-barrier) basins and the open sea. Traditionally tidal inlets have played a significant role in the regional socio-economic conditions and in recent decades the new findings about their rich and diverse ecosystems also added to their importance. Therefore there are many management issues on how to maintain, modify and benefit from these systems. In order to address such issues a fundamental understanding of the morphological behavior of these systems is necessary and has been subject of a significant amount of research. De Swart and Zimmerman (2009) summarized the research which has been carried out in this field in the last century.

Escoffier (1940) suggests that a stable tidal inlet is maintained by tidal currents. In principle the tidal currents are determined by tidal prism which is in turn determined by the combination of tidal basin geometry and tidal wave propagation into the basin. The tidal prism is also related to the size of inlet (O'Brien 1931, 1969) as well as the volume of the ebb-tidal delta (Walton and Adams, 1976). The geometry of the inlet gorge and ebb-tidal delta is a function of the ratio of wave and tidal energy (e.g. Hayes, 1975; Oertel, 1975; Hayes, 1979; Davis and Hayes, 1984). Wave energy tends to move sediment shoreward, therefore wave-dominated ebb-tidal deltas are pushed close to the inlet throat, while tide-dominated ebb-tidal deltas extend offshore. Sha and Van den Berg (1993) discuss the role of waves on the direction of the ebb-tidal delta asymmetry. They argue that the main effect of the waves is the net longshore drift, caused by wave-induced longshore current. This longshore drift is able to push the ebb-tidal delta in the direction of the wave-induced current; however this effect is depending on the relative importance of the tidal currents and the wave-induced longshore current.

Bruun and Gerritsen (1959) showed the importance of sediment by-passing. In tidal inlet systems sediment by-passing can be defined as the sediment transport from the up-drift coast to down-drift coast over the ebb-tidal delta. They interpreted the effect of wave and tide to different sediment bypassing mechanisms. They categorized the natural sediment bypassing mechanism with the ratio between longshore sediment transport by waves and tidal inlet currents. If the ratio between longshore sediment transport by waves and tidal inlet currents is high or in other words, the wave action dominates, the mechanism of sediment bypassing is called 'Bar bypassing'. In this mechanism a sand bar is formed in front of the inlet. The wave generated longshore sediment transport occurs over the bar in the direction of the down-drift coast. The depth of the submerged bar is restricted by the breaker depth of the waves. If the longshore sediment transport increases the submerged bar will get shallower and wider. In contrast, if the ratio between the longshore sediment transport by waves and the tidal inlet currents is low, bypassing occurs by tidal flow action. In this mechanism which is called 'tidal flow bypassing' sediment transports through the channels. Since the material is deposited on the up-drift bank of the channels, tidal channels migrate to down-drift. The bars between the channels may also follow this migration and join the down-drift barrier coast. The most recent conceptual model for sediment bypassing are the models proposed by FitzGerald et al. (2000) to explain the sediment bypassing in mixed-energy tidal inlets. Later these models were shown to be valid in a wide range of mixed-energy tidal dominated inlets (Elias 2006). These models are illustrated in Figure 5-1c.

Clearly the ratio between wave and tidal energy and consequently ratio between (longshore) sediment transport by waves and tidal currents is one of the governing parameters in tidal inlet morphology. Therefore tidal inlets can be classified on this basis. Oertel (1975) interpreted this ratio to the ratio of the forces of longshore and cross-shore currents and introduced a classification for ebb-tidal delta morphology (Figure 5-1a). Hayes (1979) also used this principle to classify the tidal basins based on wave height and tidal range. Later Davis and Hayes (1984) showed that tidal prism is more important than tidal range, and modified Hayes (1979) classification. Figure 5-1b shows the Davis and Hayes classification.

Using the available data about the characteristics of tidal inlet systems a few empirical relations are developed; among these relations the most well known ones are such as the relation between tidal prism and inlet cross-sectional area (O'Brien 1931, O'Brien 1939, Jarrett 1976), tidal prism and volume of ebb-tidal delta (Walton and Adams ,1976), and the tidal prism and volume of channels (Eysink, 1992).

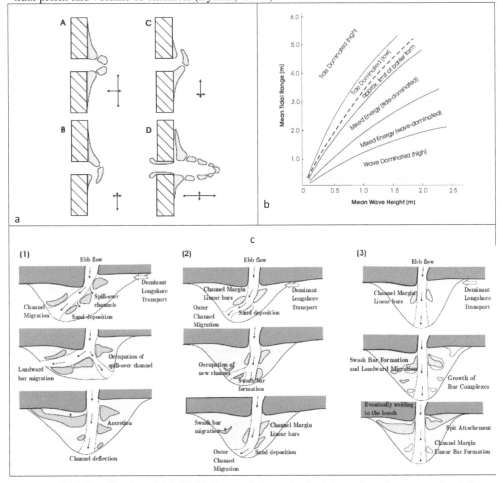

Figure 5 - 1 (a) Classification of ebb-tidal delta morphology after Oertel (1975); vector represents relative forces of cross-shore and longshore currents and (b) Hydrodynamic classification of tidal inlet after Davis and Hayes (1984) c. Bypassing model of FitzGerald et al. (2000) as presented in Elias (2006)

Most of the above mentioned descriptions, conceptual models and empirical relations are based on physical observations and empirical data; however in recent years advances in the knowledge of numerical modeling of the physical processes together with technological developments made it possible to use process-based models for mid- and long-term morphological simulations, and to study the morphological behavior of complex coastal systems such as tidal inlets. A process-based model making use of a a 'realistic analogue' approach (Roelvink and Reniers, 2012) can identify the responsible processes in morphological changes of tidal inlet systems and give more insight into the behavior of these systems. This approach has been used by Van Leeuwen at al. (2003), Van der Wegen and Roelvink (2008), Van der Wegen et. al. (2008), Dissanayake et. al. (2009), mainly to simulate the evolution of a tidal inlet system during the time under different initial conditions and forcing. They have compared their results with existing conceptual models and empirical relations. Van der Wegen et. al. (2010) tried to clarify the tidal prism - cross-sectional area relation by simulating the morphological evolution of a tidal embayment over millennia under only tidal forces.

Tung et al. (2008, 2009) used process-based morphodynamic simulations to investigate tidal inlet cross-sectional stability under different conditions of waves and tides. Those simulations showed a fair agreement between model results and the Bruun et al. (1978) empirical criterion for location stability. Also he could reproduce the Escoffier (1940) stability diagram using the outcome of the simulations (Tung , 2011 Chapter 4). Finally, Nahon et. al. (2012) carried out morphological simulations for a small schematized tidal inlet system with a 7 km^2 basin and a 700 m long and 300 m wide inlet under various wave and tide conditions. In those simulations the wave height and tidal range are systematically changed to produce different energy regime for different simulations. The outcome of those simulations is in good agreement with tidal prism-cross-sectional area empirical relation of O'Brien (1931, 1969) and resembles the available conceptual models.

5.2 Aim of the study

Despite all the above-mentioned studies a more systematic analysis of the dynamics of ebb-channel and ebb-tidal delta under different wave-tide conditions is missing. Therefore the main objective of this study is to investigate the effect of the ratio of the tidal energy and the wave energy and related processes on the dynamics of ebb-channel and ebb-tidal delta in a tidal inlet system.

To fulfill the main goal of this study we have made use of a process-based model with a setup very similar to the model used by Dissanayake et al. (2009b), which resembles the overall geometry of the Ameland inlet in the Dutch Wadden Sea. Similar to Nahon et. al. (2012), we systematically changed the tidal amplitude and the wave height in the model and carried out a series of long-term morphological simulations. In these simulations we have used a very schematized bathymetry as initial condition and we let the model develop the ebb-tidal delta and tidal channels in the system. We have analyzed the results by means of calculating tidal prism, sediment transport, sediment by-passing, volume of ebb-tidal delta, and other parameters. Also we have looked at the dynamics of the resulting ebb-tidal delta and related them to wave-tide dominancy of the system and compared them with the empirical relations and conceptual models.

5.3 Model Description

In this study we have coupled two separate numerical modules: one to calculate the water level, flow field, sediment transport and bed level changes (FLOW) and the second one to determine the wave characteristics such as wave height, spectral period, wave direction, etc. (WAVE)

5.3.1 FLOW module

For the FLOW module we have used a depth averaged (2DH) version of the flow module of Delft3D model. This model is extensively described by Lesser et. al. (2004) and Van der Wegen and Roelvink (2008). The model uses a finite difference-scheme, which solves the momentum equation including the wave generated forces (F_x and F_y) and continuity equations on a curvilinear grid with a robust drying and flooding scheme:

$$\frac{\partial \bar{u}}{\partial t} + \bar{u}\frac{\partial \bar{u}}{\partial x} + \bar{v}\frac{\partial \bar{u}}{\partial y} + g\frac{\partial \zeta}{\partial x} + c_f \frac{\bar{u}\left|\sqrt{\bar{u}^2 + \bar{v}^2}\right|}{h} - \nu\left(\frac{\partial^2 \bar{u}}{\partial x^2} + \frac{\partial^2 \bar{u}}{\partial y^2}\right) - f_{cor}v - \frac{F_x}{\rho h} = 0 \qquad (1)$$

$$\frac{\partial \bar{v}}{\partial t} + \bar{v}\frac{\partial \bar{v}}{\partial x} + \bar{u}\frac{\partial \bar{v}}{\partial y} + g\frac{\partial \zeta}{\partial y} + c_f \frac{\bar{v}\left|\sqrt{\bar{u}^2 + \bar{v}^2}\right|}{h} - \nu\left(\frac{\partial^2 \bar{v}}{\partial x^2} + \frac{\partial^2 \bar{v}}{\partial y^2}\right) + f_{cor}u - \frac{F_y}{\rho h} = 0 \qquad (2)$$

$$\frac{\partial \zeta}{\partial t} + \frac{\partial[h\bar{u}]}{\partial x} + \frac{\partial[h\bar{v}]}{\partial y} = 0 \qquad (3)$$

The velocity field obtained by solving the equation of continuity and the momentum equations is used to calculate the sediment transport field. Every time step, as a consequence of the divergence of the sediment transport field, the bed level is updated. No explicit additional process for cross-shore sediment transport is added to the model. The flow, sediment transport and bed-level updating run with the same (small) time steps. Since the morphological changes are calculated simultaneously with the other modules, coupling errors are minimized. But because this approach does not consider the difference between the flow and morphological time step, a 'morphological factor' has been applied to increase the rate of depth changes by a constant factor in each hydrodynamic time step (Lesser et al., 2004, Roelvink, 2006).

5.3.2 WAVE module

For the WAVE module we have used the stationary version of the wave solver of the XBeach model. This solver solves the wave action balance equation with a limited number of physical processes (Roelvink et al., 2009). The XBeach module considers a mean frequency in the directional space, and the spectral evolution is described by : local rate of change of action density in time, propagation of action in space with celerity of C_x and C_y, depth and current-induced refraction and spatial dissipation of wave energy due to breaking. The wave action balance in its complete form is shown here (In this study the stationary mode is used, therefore the first term is omitted from the equation):

$$\frac{\partial A}{\partial t} + \frac{\partial C_x A}{\partial x} + \frac{\partial C_y A}{\partial y} + \frac{\partial C_\theta A}{\partial \theta} = -\frac{D}{\sigma} \qquad (4)$$

Where the action density is (S_w, being the wave energy):

$$A(\theta) = \frac{S_w(\theta)}{\sigma} \tag{5}$$

This solver also includes the roller energy equation by including the dissipation of wave energy from the action balance as the source term. Therefore, the wave forces are determined by the wave induced and roller induced radiation stress tensors as described by Roelvink et. al. (2009, 2010).

5.3.3 Coupling the Modules

Initially, the WAVE module runs using the initial conditions of bathymetry, water level, etc and forcing boundary conditions. Then the resulting wave characteristics are communicated to the FLOW module. FLOW module includes the wave characteristics mainly in the form of wave generated forces in the flow computation which runs for a given number of time steps. After the FLOW computation is finished for these time steps, the updated bathymetry, water levels, and flow field are fed back to the WAVE module. In theory the exchange of data can occur at every time step of the FLOW computation but the changes of the water level, flow field and bathymetry in one time step usually are very small and also the wave conditions are not changing in such a short time, therefore the waves are updated at a specified frequency. The updated bathymetry is used in the subsequent wave computation. Figure 5-2 shows the scheme of coupling WAVE and FLOW.

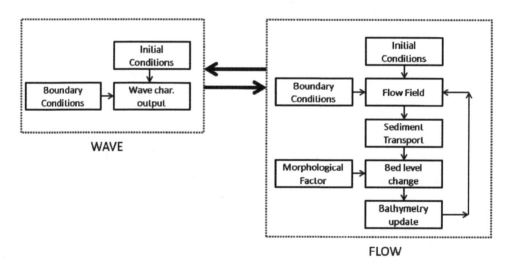

Figure 5 - 2 Scheme of coupling WAVE and FLOW

5.4   Model Setup

5.4.1   Grid and Bathymetry

The model which is used in this study is an adaptation from the schematized model of Ameland tidal basin in Dutch Wadden Sea used by Dissanayake et al. (2009b). In order to get more detailed results in the inlet, the computational grid is 3 times finer and adopted for the XBeach wave module, consequently the bathymetry is slightly different. The grid and the bathymetry of this model are shown in Figure 5-3.

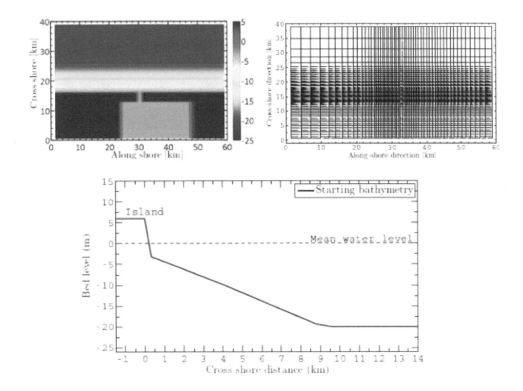

Figure 5 - 3 Grid (every 3rd grid is shown) and bathymetry of the model and a cross-shore section

As it is shown in the Figure 5-3, the initial bathymetry of the model is schematized. The bathymetry consists of a 25 x 15 km basin with constant depth of -3. This basin is connected to the open sea via an inlet 3 km long and 1 km wide. The banks of the inlet are erodible therefore the inlet can expand in width. At the sea side the seabed profile is a concave profile from 0 m to -20 m depth.

5.4.2   Sediment and sediment transport relation

In this research, non-cohesive, sediment with a median grain diameter (D50) of 200 µm is used. Therefore the sorting and armoring effects are systematically ignored. All over the model domain 40 m of sediment is available. The sediment transport relation used in the simulations of this study is based on the Soulsby - Van Rijn relation described by Soulsby (1997). This relation is a semi-empirical relation for the sediment transport in combined wave and current flow field. In this study, we have used this transport relation because it has a relatively simple formulation and the relative influence of wave and current on the total sediment transport can easily be isolated and analyzed separately.

In this relation, the bed-load transport is calculated by

$$\vec{q_b} = A_{cal} A_{sb} \vec{U_E} \xi \tag{6}$$

where $A_{cal} = 0.7$ is a calibration coefficient, A_{sb} is the bed-load multiplication factor, and ξ is a factor that represents the stirring of sediment by the waves and the current. Thus, the total bed-load transport is proportional to the current multiplied with the sediment that has been stirred up. The bed-load multiplication factor is given by

$$A_{sb} = 0.05h \left(\frac{D_{50}/h}{\Delta g D_{50}} \right)^{1.2} \tag{7}$$

with h is the water depth; Δ is the sediment relative density and g is the gravitational constant. The suspended load transport q_s is calculated by

$$\vec{q_s} = \vec{U} h c \tag{8}$$

where $\vec{U}$ is the depth-averaged water velocity, h is the water depth, and c is the depth-averaged suspended sediment concentration. This last term is calculated by solving the advection-diffusion (mass-balance) equation. In depth-averaged simulations, the 3D advection-diffusion equation as described by Lesser et al. (2004) is approximated by the depth-integrated advection-diffusion equation

$$\frac{\partial(hc)}{\partial t} + U \frac{\partial(hc)}{\partial x} + V \frac{\partial(hc)}{\partial y} - D_H \left[\frac{\partial^2(hc)}{\partial x^2} + \frac{\partial^2(hc)}{\partial y^2} \right] = h \frac{(c_{eq} - c)}{T} \tag{9}$$

Here, D_H is the horizontal sediment diffusion coefficient, and h is the local water depth. The terms on the left-hand side represents the local time evolution, the horizontal advection, and horizontal diffusion of the depth-averaged sediment concentration. The right-hand side represents erosion and deposition of suspended sediment. It depends on the depth-averaged equilibrium concentration, c_{eq}, and the sediment concentration adaptation time-scale to the entrainment of the sediment T (Galappatti, 1983). This time-scale depends on the settling velocity of the suspended sediment. The equilibrium concentration is given by

$$c_{eq} = \frac{1}{h} A_{cal} A_{ss} \xi \tag{10}$$

where A_{ss} is the suspended load multiplication factor, A_{cal} is the calibration coefficient, and ξ is the stirring factor.

In this sediment transport relation, one of the key factors is the sediment stirring term ξ. In this study, a general form of this factor is used, which allows us to change the role of waves in this factor and subsequently in the total sediment transport. The generalized form of ξ reads :

$$\xi = \left[\left(|\overrightarrow{U_E}|^2 + \frac{0.018\alpha}{C_D} U_{rms}^2 \right)^{1/2} - U_{cr} \right]^{2.4} \tag{11}$$

where $U_{rms} = \sqrt{2}U_{orb}$ is the root-mean-square near bottom wave orbital velocity, C_D is the drag coefficient due to current alone, and U_{cr} is the threshold velocity that is needed to get sediment in motion. We added the new parameter of α. The parameter α is used to change the relative importance of the (root mean square of the) wave orbital velocity (U_{rms}) to the Eulerian depth averaged velocity ($\overrightarrow{U_E}$). Note that not all wave influences are affected by α. The depth averaged velocity still contains the wave-induced longshore current and is subject to enhanced friction due to the presence of the waves. Changing α is, thus, not equivalent to changing all the wave characteristics.

Effectively α changes the calibration coefficient of 0.018. This coefficient is derived from calibrating this sediment transport formulation with the Van Rijn transport relation (Soulsby, 1997). It can be argued, however, that this constant is likely to differ from its current value since it was calibrated on measurements that were done in the shoaling domain with water depths in the order of 5 m and no breaking waves (Soulsby, 1997). It is not straightforward to imply that the resulting calibration coefficient is representative for shallow coastal regions where waves break.

5.4.3   Forcing of the model

To fulfill the goal of this research and compare the morphological characteristics of tidal inlet systems under different energy regimes, we used the classification of Davis and Hayes (1984) to determine the corresponding wave height and tidal range for each energy regime. The tidal forcing used in this study, is a simplification of the tidal configuration at Ameland inlet in Dutch Wadden Sea. Following Van de Kreeke and Robaczewska (1993), the spring neap cycle is ignored and the dominant forcing by M2 and over-tides (M4 and M6) is considered. This tidal forcing is applied in the model as a water level boundary at the offshore boundary of the model and Neumann boundaries, where the alongshore water level gradient is prescribed, for lateral boundaries of the model (Roelvink and Walstra, 2004). Details of the tidal forcing are presented in Table 5-1. These tidal components lead to a mean tidal range of approximately 1.7 m at the inlet. In order to carry out simulations with different values of the tidal range corresponding to different energy regimes, all amplitudes of the tidal components in Table 5-1 are multiplied by a constant factor (0.5, 1.0 and 1.5). For the waves also we have used the representative wave condition for the same area of Dutch coast; this representative wave is northwesterly wave (330 Deg) with the wave height of 1.4 m (Dissanayake 2011). To move in the energy domain we simply multiply this wave height with 0.5, 1.0 and 1.5 for different cases.

Table 5 - 1 Tidal constituents at the boundaries of the model without multiplication factor (Based on Dissanayake et al. 2009)

Tidal component	Frequency (deg/hr)	West		East	
		Amplitude (m)	Phase (deg)	Amplitude (m)	Phase (deg)
M2	28.9933	0.8450	20.2	0.9200	53.3
M4	57.9866	0.0938	259.5	0.0861	304.4
M6	86.9799	0.0625	119.5	0.0409	225.1

5.5   Different simulations

Based on the abovementioned forces including 3 different wave heights and 3 tidal ranges plus 3 case without wave, 11 different cases has been determined (Due to too much morphological activity the case with highest tidal range and highest wave height is omitted). In Figure 5-4 different forcing conditions for all the simulations are shown on the energy regime classification of Davis and Hayes (1984). Nahon et. al. (2012) also used the same classification to determine the forcing of their idealized model, however they have carried out short time morphological simulations for a small tidal inlet system and they analyzed the outcome of the models only in terms of of tidal prism/inlet cross-sectional area.

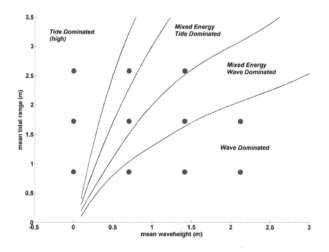

Figure 5 - 4 Forcing conditions for different simulations on classification of Davis and Hayes (1984)

5.6   Results and Dissuasions

5.6.1   Morphology of ebb-tidal delta

The resulting bathymetry of different cases shows that the morphology of the ebb-tidal delta as it is expected is highly dependent on the relative energy of tides and waves. Figure 5-5 shows the resulting bathymetry of all the cases after 20 years of morphological simulation.

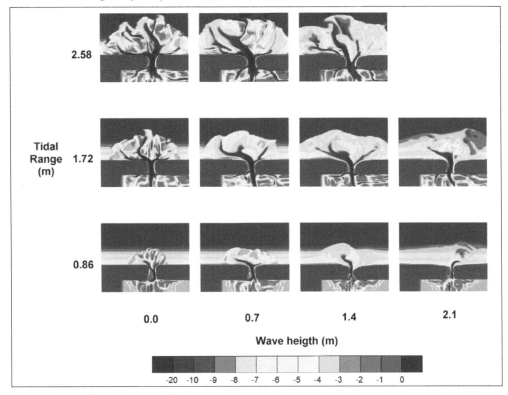

Figure 5 - 5 Resulting bathymetry after 20 years of morphological simulation

In these simulations the tidal wave is propagating from west to east. In the tide dominated cases the main ebb-channel is rotating towards the up-drift coast, and the ebb-tidal delta is extending more seaward. This rotation is less pronounced in the case of larger tidal range; this is due to the fact that the higher tidal range leads to stronger cross-shore ebb-flow perpendicular to the coastline. In the cases with more wave action the offshore extension of the delta is reduced. In the most wave-dominated case the effect of the waves is so extreme that the orientation of the channel at this specific time is changed toward the down-drift coastline; however it should be mentioned that in the wave-dominated cases there is a cyclic behavior in the channel orientation which is discussed in the next section. Also higher waves push more sediment towards the coastline and this leads to reduced sand volume of ebb-tidal delta. The same features can be seen in Oertel (1975) classification as well as the conceptual models of Sha and van den Berg (1993).

The size of an ebb-tidal delta is measured by its total sand volume. Walton and Adams, (1976) used the available data to develop a relation which relates this volume directly to the mean tidal prism of the corresponding tidal (back-barrier) basin:

$$V = cP^{1.23} \tag{12}$$

Where V is the volume of the ebb-tidal delta, and P is the mean tidal prism, c in the equation depends on the wave energy. The value of $6.6e^{-3}$ is suggested as an average value for c (Dean, 1988; Walton and Adams, 1976). Dean (1988) shows that, for constant tidal prism, total volume of the ebb-tidal delta will decrease when it is exposed to higher wave energy. This is because there is more available wave energy to drive the sand back to shore in high energy environments.

Figure 5-6 shows the changes of the ebb-tidal volume, calculated based on Walton and Adams (1976), for different cases. As expected the simulations with larger tidal range produce larger ebb-tidal deltas. But, as it is clear in this figure, in simulations with lower tidal range, interestingly adding wave action initially increases the volume of ebb-tidal delta but with larger waves this volume reduces significantly. This is due to the fact that initially adding wave increases the bed shear stress and more sediment is stirred in the water and more sediment is available to form the ebb-tidal delta and the presence of the small wave keeps the sediment close to the inlet and in the area of ebb-tidal delta. But when the wave height increases the wave induced current erode the forming ebb-tidal delta and transport the sediment towards the down-drift coast. In the case with the largest tidal range since the share of wave in the resulting bed shear stress and consequent sediment stirring is not significant, the initial increase of ebb-tidal delta volume by adding waves does not exist and adding waves reduces the ebb-tidal volume.

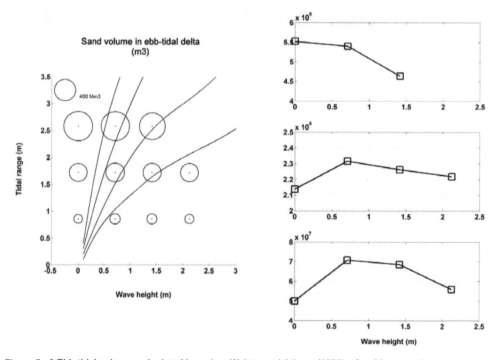

Figure 5 - 6 Ebb-tidal volume, calculated based on Walton and Adams (1976), after 20 years of morphological simulation

In Figure 5-7, the relation between tidal prism and the ebb-tidal volume for all the 11 simulations after 10 (gray circles), 20 (gray squares), 30 (black circles) and 40 (black squares) years of morphological simulation is shown and the general relation of tidal prism and the ebb-tidal volume (Equation 12) is fitted to all the data of each year to find the corresponding coefficient of c. The fitting lines are also shown in Figure 5-7. This figure shows that in the cases with higher tidal range the effect of different wave heights on this relation is reduced. Also it is obvious in this figure that the value of c increases with time and gets closer to the metrical value of $6.6e^{-3}$. Van der Wegen (2010) showed the same time dependency for the relation between tidal prism and cross-sectional area.

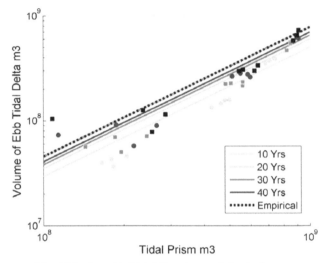

Figure 5 - 7 Changes of the tidal prism ebb-tidal delta volume relation in time

5.6.2   Sediment bypassing and cyclic behavior of main channel

Sediment bypassing is the key mechanism governing the inlet morphology and the effect of the inlet on the adjacent coastlines. To investigate sediment bypassing and cyclic behavior of main channel we have chosen 5 of the abovementioned 11 simulations. The wave height and the tidal range for these runs together with their energy regime is summarized in Table 5-2.

Table 5 - 2 Simulations considered investigating Sediment bypassing

Energy regime	Tidal range (m)	Wave height (m)
wave dominated (high)	0.86	2.1
wave dominated	1.72	2.1
Mixed energy, wave dominated	1.72	1.4
Tide dominated (low)	2.58	0.7
Tide dominated (high)	1.72	0.0

As it is shown in Table 5-2, each of theses simulations belongs to a different energy regime; here we have tried to check the condition of each simulation with respect to the conceptual bypassing model of Fitzgerald et al. (2000).

The evolution of tidal inlet system in the highly wave dominated regime is shown in Figure 5-8. In this simulation we can observe the cyclic ebb-tidal delta breaching; initially the inlet is generated in about 6 years and then, due to sediment accumulation at the up-drift side of the ebb-delta, the main ebb channel migrates towards the down-drift coast. These shoals deflect the main ebb-channel in down-drift direction and due to the shoreward component of the horizontal circulation behind these shoals, some of these shoals land at the down-drift coast. But at this time the curved channel is hydraulically less efficient; therefore at year 10 the ebb-channel breaches the ebb-tidal delta and a more competent, seaward-directed channel through the ebb-tidal delta forms. The same cycle of ebb-channel migration to the down-drift coast and seaward breach of ebb-tidal delta repeats two more times in the 50 years duration of the simulation, the time scale of each cycle is about 15-20 years. It is also clearly shown that when the volume of sediment in the ebb-tidal delta is larger, the morphological cycle is longer. This type of cyclic behavior corresponds with the first type in the conceptual model of FitzGerald et al. (2000) shown in Panel C of Figure 5-1.

Moving from the highly wave dominated to the wave dominated (Figure 5-9) and wave dominated mixed energy (Figure 5-10) regimes we can see a similar behavior as in the previous condition, but the cyclic channel migration is limited to the outer part of the channel while the inner part remains stable and the main channel always bends toward the up-drift coast, which is due to the direction of tidal wave propagation. This behavior corresponds with the second type in conceptual model of FitzGerald et al. (2000) shown in Panel C of Figure 5-1. Also the abandoned part of the outer channel migrates towards the offshore and in wave dominated case disappears in the deeper areas. There are, on the other hand, differences between these two simulations. In the wave dominated regime (Figure 5-9) we can see another cyclic behavior; a secondary channel is developed at the east side of the channel and silted up due to the shoal migration and landing on the down-drift coast. In the wave dominated mixed energy regime (Figure 5-10) the outer channel shifting is more limited and happens at a larger distance from the inlet; the secondary channel is more stable and gets larger to the extent that after 34 years the inlet has two channels with a bar in between.

In tide dominated regimes (Figures 11 and 12), the ebb-tidal delta and the main ebb-channel remain stable and bend towards the up-drift coast. Bar complexes are formed and grow offshore, in the case with waves (Figures 11) the offshore growth of the bar is limited and some bars migrate onshore due to the dominance of landward flow created by waves.

The above-mentioned simulations reproduce and explain the different conditions of the conceptual model of FitzGerald (2000), and relate them to the energy classification of Davis and Hayes (1984). Although the inlets in the same regime show similar behavior, the observed differences suggest that boundaries between the energy classes of Davis and Hayes (1984) are not clear cut. For example, in the highly wave dominated cases the cyclic behavior and ebb-tidal delta is clearly present; however with the increasing of tidal range while still remaining in the wave dominated regime, the cyclic behavior of the main ebb-channel takes place at a larger distance from the inlet, and inside the inlet a cyclic secondary channel develops. In the mixed energy regime, only at the far end of the main ebb-channel we can observe a changing in the orientation of the channel. Therefore as FitzGerald (1996) also indicates, the dynamics of tidal inlets at decadal time scales cannot be described in sufficient detail by a general classification based on wave and tidal energy. There are many parameters and processes that influence the inlet behavior, such as available sediment, basin geometry and geology and freshwater inflow when present.

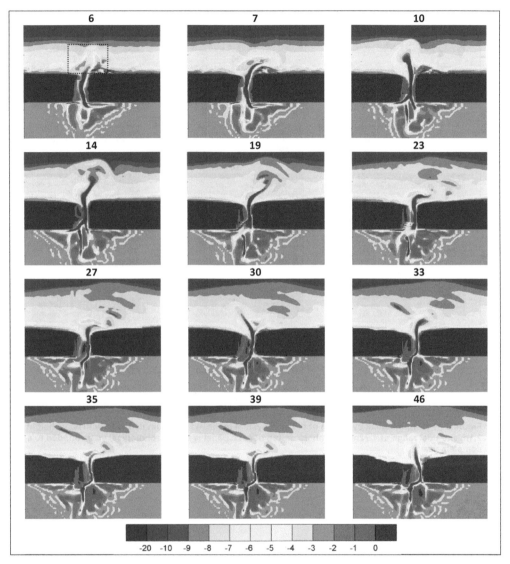

Figure 5 - 8 Morphological evolution of the inlet system in the extremely wave dominated case with wave height of 2.1m and tidal range of 0.86m. (Numbers on top of each panel is the morphological year rounded to a complete year)

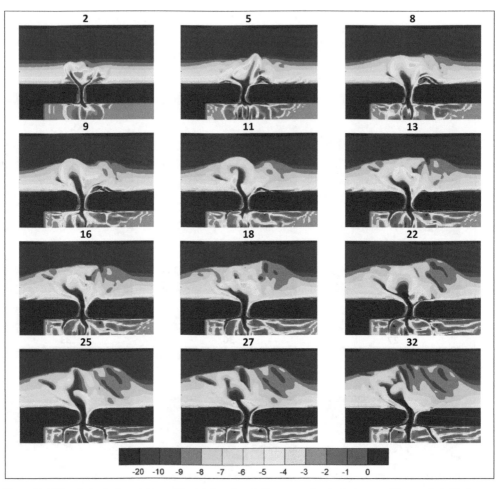

Figure 5 - 9 Morphological evolution of the inlet system in the wave dominated case with wave height of 2.1m and tidal range of 1.72m. (Numbers on top of each panel is the morphological year rounded to a complete year)

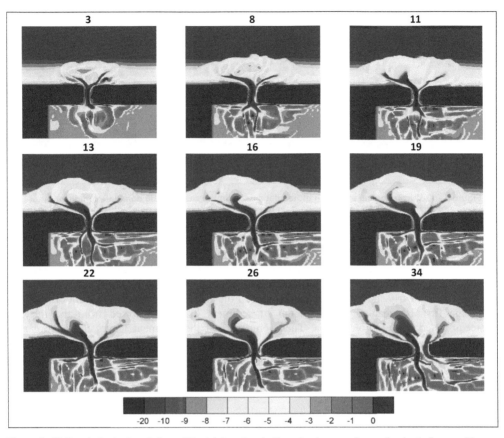

Figure 5 - 10 Morphological evolution of the inlet system in the mixed energy (wave dominated)case with wave height of 1.4m and tidal range of 1.72 m. (Numbers on top of each panel is the morphological year rounded to a complete year)

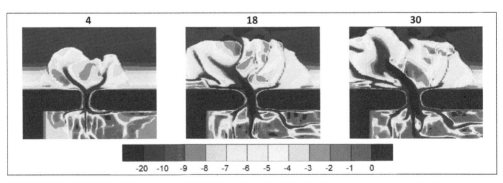

Figure 5 - 11 Morphological evolution of the inlet system in the case of tide dominated (low) with wave height of 0.7m and tidal range of 2.58m. (Numbers on top of each panel is the morphological year rounded to a complete year)

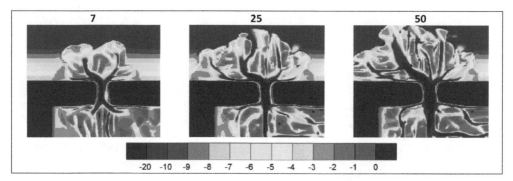

Figure 5 - 12 Morphological evolution ofthe inlet system in the case of Tide dominated (high) without Wave and tidal range of 1.72 m. (Numbers on top of each panel is the morphological year rounded to a complete year)

Also in the case of the highly wave dominated regime, we established a control area in front of the inlet (shown in the first panel of Figure 5-8). At the boundary of this control volume we have recorded the total cumulative sediment transport throughout the 50 years of simulation and calculated the changes of the sediment inside the control area (Figure 5-13). This figure clearly shows the cyclic behavior of the sediment by passing in front of the inlet, in the first 10 years a large amount of sediment, mainly exported from the inlet, is accumulated in the control area. Then some part of the sediment leaves the control area towards the down-drift coast. From this point onward the control area in front of the inlet acts as a buffer area and sediment coming from the up-drift coast and inlet is stored there for a while and then moves toward the down-drift coast.

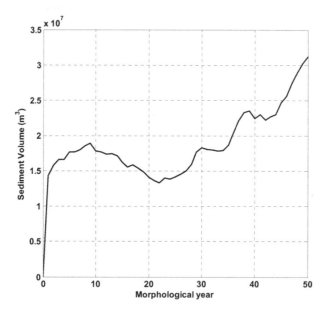

Figure 5 - 13 Change of the volume of sediment in the control area in front of inlet in the case of the highly wave dominated regime

5.6.3 Hydrodynamic and sediment transport patterns

To show the hydrodynamic and sediment transport patterns for different energy regimes, we have chosen 3 different conditions: highly wave dominated with the ebb-channel toward the down-drift coast (after 23 year in Figure 5-8); highly wave dominated with the ebb-channel extending seaward (after 30 years in Figure 5-8); and highly tide dominated with a stable channel (after 25 year in Figure 5-12). For these conditions separate hydrodynamic/sediment transport simulations were carried out and the residual flow and sediment transport patterns were computed as shown in Figure 5-14.

It is shown that in the highly wave dominated regime with the ebb-channel bending toward the down-drift coast (Figure 5-14 upper panel) the residual flow and sediment transport has different distinctive areas. The first one is an east-ward long-shore current parallel to the up-drift coast; this residual current is mainly due to the strong wave driven current combined with flood flow. This current also brings sediment and pushes the ebb-channel towards the east. The second area is the ebb-channel itself; the residual current in the ebb-channel shows that the basin is exporting sediment and part of this exported sediment lands on the south of the ebb-tidal delta. The third zone is the area of ebb-tidal delta, in this area a combination of the ebb-flow and the wave force on the north side of ebb-tidal delta generates a relatively strong residual current over the ebb-tidal delta; the sediment transport caused by this current is the main reason of bar migration towards the down-drift coast ("Bar bypassing"). In this case the shallow area between the ebb-channel and the down-drift coast is protected from waves by the ebb-tidal delta; therefore the main current here is the flood flow towards the inlet. Another interesting morphological aspect of this condition is the separation point the down-drift coast, at this coast line there is a point where the direction of both residual flow and residual sediment transport changes from west to east. This point is the border of the area protected against the waves by the ebb-tidal delta.

The second case has the same energy regime as the above-mentioned case but now the breach in the ebb-tidal delta has occurred and the seaward-directed channel through the ebb-tidal delta is formed (Figure 5-14 middle panel). In this case the major residual flows are the east-ward long-shore current parallel to the up-drift coast and the ebb-flow seaward jet. Most of the sediment carried by the long-shore current is flushed out by the ebb-flow and deposited at the far end of the ebb-channel, gradually causing the channel to rotate eastward. In this case the flood residual flow exists only on the shallower areas of the inlet.

In the last case with highly tide dominated energy regime (Figure 5-14 lower panel), the ebb-tidal delta and ebb and flood channels are fully developed and the residual currents are mainly in these channels. The residual sediment transport in the channels is negligible. This is the main reason of the stability of the channels. The residual flow and sediment transport on the shoals only reshape the shoals but, the overall shape of the delta remains the same.

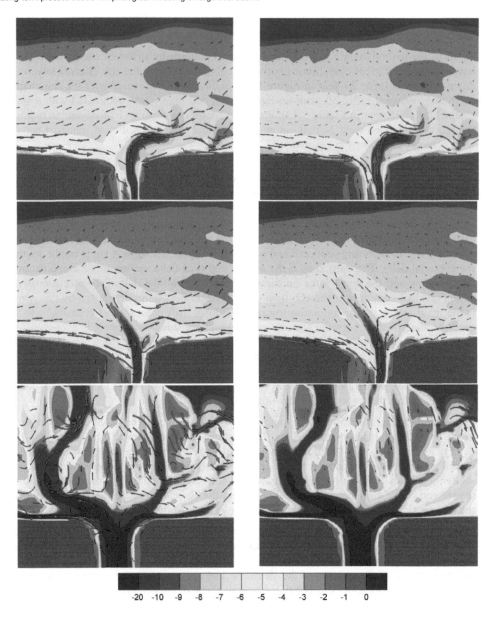

Figure 5 - 14 Residual flow (left panels) and residual sediment transport (right panels) fields for highly wave dominated (upper and middle panels) and highly tide dominated (lower panel) energy regimes.

5.6.4   Effect of wave stirring

As mentioned before, in this study we have used a general form of the stirring term ξ, in Soulsby - Van Rijn sediment transport relation. In this section, we have changed the importance of the wave and current in the stirring term by changing α in the equation (11).

Figure 5-15 shows the ebb-tidal delta for the mixed energy (wave dominated) regime for different values of α after 30 years of morphological simulation. It is shown that in the case of $\alpha=0$, the developed ebb-tidal delta is more similar to tide dominated ebb-deltas, with stable channels, shallower bars and more offshore extension. By increasing the value of α the wave dominated features appears in the ebb-tidal delta. The volume of ebb-tidal delta shows a small decay by changing α from 0.0 to 0.3 but later it increases significantly, this trend is the exact opposite of the trend which was observed by adding waves to simulation (Figure 5-6). These two different trends can be explained as follows. Waves introduce two main processes into the model: wave induced current and additional bed shear stress. A zero value for α only omits the share of the wave orbital velocity in the bed shear stress and not the wave induced current. Therefore from Figure 5-15 it can be concluded that the role that waves play in increasing the bed shear stress has a more significant role in the wave dominated behavior than the wave induced currents.

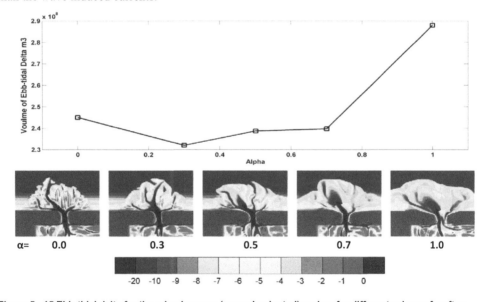

Figure 5 - 15 Ebb-tidal delta for the mixed energy (wave dominated) regime for different values of α after 30 years of morphological simulation

5.7   Conclusion

In this study we compared the outcome of morphodynamic simulations with conceptual models and empirical relations in the case of a medium size tidal basin with different wave and tide conditions. In view of the simplicity and ease of reproduction we chose to carry out these simulations with the Soulsby-van Rijn formula, which gave qualitatively realistic and robust results. A good agreement between the morphological models and conceptual models on a decadal time scale is observed in this study. Therefore we can conclude that this type of simulation can give more insight into the processes involved in the morphological behavior of tidal inlets. We could simulate the cyclic behavior of the ebb-channel in the wave dominated regime and show that this behavior can be explained independent of the changing wave direction and without explicitly considering cross-shore transport effects, e.g. due to wave skewness and asymmetry.Our simulations support the Fitzgerald et al. (2000) conceptual model of the behavior of ebb-tidal delta and show qualitatively similar behavior for tidal inlets in the same energy regime based on Davis and Hayes (1984). However the distinctions between different energy regimes are not completely straightforward. We have shown that the result of the model depends on the formulation of the wave current interaction effect on the sediment transport.

CHAPTER 6

Unleashing the waves

Wave schematization approaches for long-term morphological modeling of tidal basins

6.1   Introduction

Since 1990, the Netherlands has adopted the policy of "Dynamic Conservation", aiming to maintain the coastline position, and avoiding any systematic landward retreat. Additional goals of this policy concern the preservation of valuable dune areas and of the natural dynamic character of the coast (van Koningsveld and Mulder, 2004). The Dutch Waddenzee forms one the three distinct coastal systems which also includes the Delta and the Holland Coast. Considering the complex morphodynamic of the Dutch Waddenzee, the challenges in maintaining this area and protect it from drastic changing such as drowning due to sea level rise, are among the first priorities of the coastal zone managers in the Netherlands. Therefore there is a need for better understanding and assessing the qualitative and quantitative coastal developments that will occur in response to changing environmental conditions and/or human measures in shape of coastal management scenarios (Wang et. al 2011). Therefore in recent decades many different research lines have been established in the field of morphodynamics of the Dutch Waddenzee, such as large-scale sediment budget studies, short term process-based simulations of hydrodynamic and sediment transport, investigating morphodynamic behavior of smaller scale features, semi-empirical modeling of the effect see-level rise, etc (Eysink and Stive, 1989, Mulder 2000,Elias et al. 2004, Walburg 2006, de Ronde, 2008, Van Koningsveld et al., 2008, Elias et al., 2008, Elias et al. 2009, Dissanayake, 2011, Dissanayake, 2012, Elias et al. submitted) providing a wide range of knowledge about morphodynamics of the Dutch Waddenzee. However a process-based model has not been used to try to hindcast the morphological changes of the area on a large scale and for a long duration. Using a process-based model for a long-term hindcast simulation confronts us with a wide range of challenges, but the most important challenge in this regard is to schematize the input to the model in most realistic and meanwhile efficient way. This is more challenging when the important processes and forcing in the study area are numerous and interacting. Therefore the first objective of this study is to review the input reduction techniques and investigate the effect of different wave schematization approaches on long-term process-based morphological simulation of the Dutch Waddenzee and describe different methodologies for applying schematized inputs in long-term morphological simulations. The second objective is to carry out a 50 years hindcast simulation of the Dutch Waddenzee.

6.2   Area of interest and available data

The Waddenzee is located at the south east side of the North Sea, and consists of 33 tidal inlets system along approximately 500 km of The Dutch, German and Danish coastlines. The barrier islands of these tidal basin systems separated the largest tidal flat area from the North Sea (Elias, 2006). The part of the Waddenzee which is along The Netherlands coastline (Dutch Waddenzee) is shown in Figure 6-1. The ebb-tidal delta shoals in Dutch Waddenzee are relatively large while they are associated with relatively narrow and deep channels, the back barrier basins of these tidal inlet systems consist of extensive systems of branching channels, tidal flats, and salt marshes. The tidal basins in the east and west of Dutch Waddenzee differ in different aspects : in the eastern part (Ameland and Frisian) the back barrier area is shallower including large flat areas and small channels, while in the Western part, mainly in Marsdiep and Vlie, channels are much deeper and flat areas are relatively small (Elias 2006). Continuous sedimentation in the tidal basins is one of the characteristic features of Waddenzee. This sediment demand is fed by sediment supply from the barrier coastline, ebb tidal deltas and the adjacent North-Holland coastline. As it is mentioned by Elias (2006), this problem is addressed by Stive and Eysink (1989); they note that the cause of structural large sand losses from the North-Holland coastline is mainly the demand of sand in the Waddenzee tidal basins. Therefore morphological developments, stability, and changes of tidal basins and tidal inlets in Dutch Waddenzee have a huge influence on the Dutch coastline mainly in terms of extensive erosion (need for nourishment) of the barrier islands and adjacent coasts. The main area of interest in the current study is the Western part of Dutch Waddenzee.

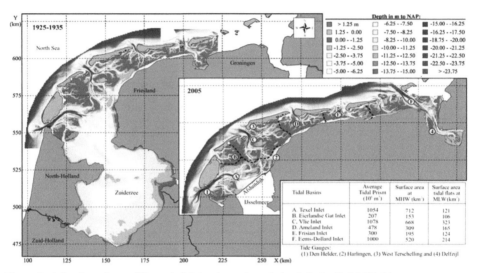

Figure 6 - 1 Configurations of the main inlets, channels and shoals in the Dutch Waddenzee representative for the barrier 1925-1935 (large figure) and 2005 (insert) (Elias et al, submitted)

6.2.1   Wave data

In general, the dominant waves affecting the Waddenzee are the wind generated waves in the North Sea while the swells have relatively minor contribution (Elias, 2006, Roskam, 1988; De Ronde et al., 1995; Wijnberg, 1995). There are nine stations for wave observation spread along the coast of the Netherlands. The data of these stations are made available by KNMI (www.watermarkt.nl). The closest wave measurement station to our area of interest is the one close to the Eierlandse Gat inlet. The wave data of this wave observation station are analyzed and used in this study. The mean significant wave height at this station is about 1.3m coming from south- west with a period of 5 seconds. However, within the observation period the wind generated waves can reach the height of 6m causing storm surges of up to 2m.

Table 6-1 summarizes the wave conditions. The wave directions have been classified into twelve directional bins where each bin covers 30 degrees; The wave height is classified into eight bins with one meter increments. The complete wave rose of the data is shown in Figure 6-2. The dominant wave directions are South West and North. The majority of the waves (about 40%) are less than one meter high while only 20% of the wave observations exceed the 2m wave height. The highest waves come from the North West direction.

Table 6 - 1 Probability of occurrence of different wave conditions

	N	NNE	ENE	E	ESE	SSE	S	SSW	WSW	W	WNW	NNW
Hs (m)	345-15	15-45	45-75	75-105	105-135	135-165	165-195	195-225	225-255	255-285	285-315	315-345
0-1	7.20	4.88	2.27	1.02	0.74	0.83	1.17	3.33	4.81	3.69	4.38	6.33
1-2	6.09	2.79	1.92	0.75	0.43	0.35	0.81	4.79	5.80	4.06	4.52	7.12
2-3	1.35	0.51	0.38	0.07	0.01	0.03	0.11	1.89	2.89	2.04	1.93	2.88
3-4	0.25	0.08	0.01	0.00	0.01	0.00	0.02	0.39	0.99	0.78	0.79	0.97
4-5	0.05	0.00	0.00	0.00	0.00	0.00	0.00	0.03	0.18	0.31	0.29	0.31
5-6	0.02	0.00	0.00	0.00	0.00	0.00	0.00	0.00	0.02	0.10	0.07	0.08
6-7	0.00	0.00	0.00	0.00	0.00	0.00	0.00	0.00	0.01	0.02	0.02	0.02
7-8	0.00	0.00	0.00	0.00	0.00	0.00	0.00	0.00	0.00	0.00	0.00	0.00

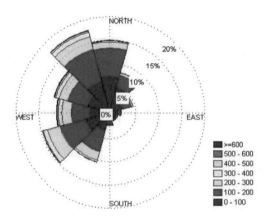

Figure 6 - 2 Wave rose for the wave data of Eierlandse Gat (cm)

Also analyzing the wave data shows that the wave climate in the area can be divided into two distinct seasons: summer lasting for six months from April till end of September with average wave height of about 1.0m and winter also for six months, from November till end of March with higher wave heights. The wave climate characteristics of the two seasons are presented as wave-roses in Figure 6-3. The effect of such seasonality on morphological change is investigated in this research as well.

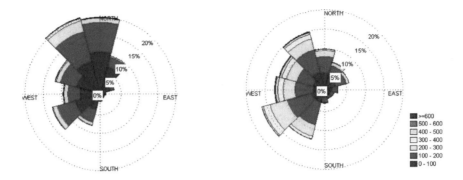

Figure 6 - 3 Wave Rose; summer (left), winter (right) (cm)

6.2.2   Tides

The vertical tidal motion along the coast of The Netherlands as well as in the Waddenzee is generated from the tidal wave from the Atlantic Ocean, entering the North-Sea from North (between Norway and Scotland) and from the south west (through the Calais Strait). These two waves interfere with each other and under the effect of Coriolis force and bottom friction, generate a complicated tidal flow pattern in the south of North-Sea. As it is shown in Figure 6-4, tides in the south of North Sea propagate from west to east rotating counter-clock wise around their amphidromic points.

Figure 6 - 4 Propagation of tide in North-Sea (www.getij.nl)

Along the Dutch coast the tide is a composition of a standing and a progressive wave, propagating from south to north. With a tidal current velocity of 0.5-1.0 m/sec this wave meets the second tidal wave near the Marsdiep inlet (Texel). The merged tidal wave moves to the east along the islands of the Waddenzee. The amplitude of the tide decreases along the Dutch coast until its minimum at Den Helder (the tidal observation point at the Marsdiep inlet) and then increases again along the Waddenzee barrier islands. At the Marsdiep inlet, the dominant tidal component is M2 and the maximum tidal current velocity in the Marsdiep is between 1.0 to 2.0 m/sec. The tidal asymmetry in Marsdiep is mainly due to distortion of the M2 by M4. The tidal range varies between 1.0 to 2.0 m with the mean value of 1.38m at Den Helder. The tidal range increases to the mean value of about 2.0m at the Vlie inlet.

6.2.3 Bathymetry

The Dutch Waddenzee is one of the best monitored coastal regions in the world. There are some depth measurements especially in Marsdiep from the 16th century. Since 1987 Rijkswaterstaat (Directorate-General of Public Works and Water Management of The Netherlands) frequently measures the bed level in the Waddenzee. The ebb-tidal deltas are measured every 3 years, while the basins are measured every 6 years. These data are stored in a 10 x 12.5 m resolution database called 'Vaklodingen'. Recently all the available data since 1926 has been collected, processed and made accessible through the OPENEARTH project (www.openearth.nl). Using this set of data it is possible to construct the historical bathymetries for different parts of the Dutch Waddenzee for different years.

6.2.4 Wind and surge

The wind and surge data used in this study was taken from one of the stations close to the Eierlandse Gat. These data cover the period of 1996 till 2001. The Wind rose is shown in Figure 6-5. The dominant wind is from the south-westerly to north-westerly directions with the average velocity of about 7 m/s.

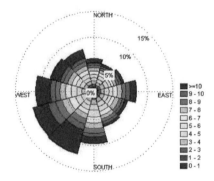

Figure 6 - 5 Wind rose (m/s)

6.3   Model Description

The model which is used in this study is a 2D (depth-averaged) version of Delft3D (Lesser et al., 2004), in which there are two numerical modules interacting with each other : FLOW and WAVE. The FLOW module is a numerical solver which is based on finite differences and solves the unsteady shallow-water equations. The system of equations consists of the horizontal momentum equations, continuity equation, transport equation and a turbulence closure model. The resulting velocity field is then fed to a sediment transport relation and then morphological changes are determined based on the mass-balance for sediment in each computational grid. Since the time scales for the hydrodynamics and morphology differ, multiplying the bed level changes in each computational grid at each time step by certain scaling-up morphological factor (MorFac) bridges the gap between these two scales. This makes it possible for long-term morphological simulations to be achieved by using hydrodynamic simulations which are just a fraction of the required duration for morphological simulations (Roelvink, 2006). In order to add the wave forcing to the shallow water equations of the FLOW module, this module interacts with the WAVE module. The Delft3D WAVE module (third generation SWAN - Simulating WAves Nearshore - model (Booij et al., 1999)) is used to simulate the evolution of random short-crested waves in coastal zones. This module interacts with the FLOW module in an online manner meaning that initially, the WAVE module runs using the initial conditions and forcing boundary conditions. Then the resulting wave generated forces are communicated to the FLOW module. FLOW module including these wave generated forces runs for a given number of time steps. After the FLOW computation is finished for these time steps, the updated bathymetry, water levels, and flow field are fed back to the WAVE module. The exchange of data between these two modules can be done every time step but if the changes of the water level, flow field and bathymetry in one time step is very small then the data can be exchanged between the modules less frequently.

6.4   Schematization of forces

6.4.1   Tide

Since we are using the morphological factor (MorFac) for the morphological simulation we have to use a schematized tide in the model. To schematize the tide, a complex time series of tidal water level and induced currents, is replaced by a simple "morphological tide". The "morphological tide" should have the same morphological effect as the real tide, i.e. it should produce the same residual sediment transport and morphological change for the duration of the simulation. The techniques used in the literature for producing representative morphological tide are based on the idea of Latteux (1995) and modified by Lesser (2009). Lesser (2009) claims that in the case of semi-diurnal tide, a M2 tide with 7-20% larger amplitude can be considered as the morphological tide. However, in shallower areas such as Waddenzee M2 tide is distorted by M4 and M6 constituent, so these two constituents should be used in producing the "morphological tide" (Chapter 2, Van de Kreeke and

Robaczewskam 1993). Based on Lesser's (2009) suggestion for "morphological tide" if the diurnal components (O1 and K1) of the tide are considered to be important in the morphological simulation, theses components can be taken into account by adding an artificial diurnal component (C1). The amplitude and phase of C1 is computed by using the following relations given that O1 and K1 are the amplitudes, and φ_{O1} and φ_{K1} are the phases of O1 and K1 respectively.

Amplitude: $\quad\quad\quad\quad\quad C_1 = \sqrt{2\,O_1 K_1}$

Phase: $\quad\quad\quad\quad\quad\quad \emptyset_{C1} = \frac{\emptyset_{O1} + \emptyset_{K1}}{2}$

The morphological tide which is used in this study is derived from the 'ZUNO' model which is a calibrated model for the vertical tide in North Sea (Roelvink et al., 2001). The model is run for one year and the tidal variation is recorded at the boundary of the Waddenzee model. These recorded tidal variations are analyzed to obtain M2, M4, M6, O1 and K1. Combination of these constituents forms the morphological tide for this study.

6.4.2 Wave

The wave climate also should be schematized to limited number of representative wave conditions as the input of the long-term process based morphological model. Lesser (2009) identified two main questions which should be answered for wave schematization:

1. At what scale does the chronology of waves become important? and
2. At scales in which the chronology of wave events can be assumed to be unimportant, what representative set of wave classes can be used to produce the same residual sediment transport and morphological change patterns over the area of interest as the full wave climate?

He mentioned that in order to have the same morphological changes, the pattern and the rate of residual sediment transport due to the whole wave climate should be the same as the schematized wave conditions, therefore these parameters can be used as the schematization criteria.

In this study we have omitted the waves coming from 45-195 Deg. (10% of the waves in total) and used three different approaches for wave schematization:

- Potential sediment transport method,
- Energy flux method,
- and "Opti" Method.

Potential sediment transport approach

In this approach the wave climate is divided manually into a number of wave classes and directional bins. the representative wave condition of each cluster of conditions is determined based on the potential of sediment transport of those selected wave conditions.

The representative wave height for each cluster is calculated using the principle of CERC formula in which the potential of sediment transport is related to the wave height to the power of 2.5. Therefore the representative wave height is calculated using the following relations. It means that the representative wave height has the same potential sediment transport as the whole wave cluster

$$H_{rep}.^{2.5} \sum P_i. = \sum P_i H_i^{2.5} \qquad (1)$$

$$H_{rep} = \left(\frac{\sum P_i H_i^{2.5}}{\sum P_i.} \right)^{1/2.5} \qquad (2)$$

In which

H_i : the significant wave height in the bin "i"

P_i : the probability of wave height in the class

The representative wave period and direction are calculated as follows:

$$Dir_{rep} = mean(dir_i) , \quad T_{rep} = mean(T_i) \qquad (3)$$

In which

dir_i : wave direction in bin i calculated based on the statistical analysis,

T_i : wave period in bin i calculated based on the statistical analysis

In total 4 different scenarios of clustering of wave conditions are used to schematize the wave climate using this method.

For the first scenario we have chosen two different clusters for calm period and storm conditions. For the waves up to 3 meters, the wave conditions are divided into three wave height classes of one meter interval and two directional bins : 195 to 285 and 285 to 45 degrees nautical . The storms with wave height of 3 meter and higher are clustered into one wave height class and three directional bins: 315 to 45, 195 to 255, and 255 to 285.

In the second scenario, the wave conditions are divided into three directional bins; 315 to 45, 195 to 255, and 255 to 315 and two wave height classes. Because about 80% of the wave observations do not exceed two meter height, this value is chosen as the border of the wave height classes. Therefore the wave climate is schematized to 6 wave conditions.

Also in the third scenario, the whole wave climate is schematized in 6 wave conditions but with only one wave height class and 6 directional bins with 30 degree intervals.

To investigate the effect of seasonality on the morphological changes of the area of interest, we decided to develop a scenario in which the wave schematization includes two distinct seasons. Each season includes six months and alternating in each year of simulation. The schematization is the same as the aforementioned scenario one but for two different seasons. All the representative wave conditions for this approach are summarized in Appendix.

Wave Energy Flux Approach

The concept of the wave energy flux is applied in this approach of wave schematization. This method was used by Dobrochnski (2009) for short term morphological modeling of a coast stretch in Brazil.

In this approach, the energy flux of each wave in the data set is calculated using the following equation:

$$E_f = (\rho.g.H_s^2 / 8).C_g \qquad (4)$$

In which

ρ : water density

g : gravitational acceleration

H_s : significant wave height

C_g : wave group velocity

Then the numbers of the directional bins and wave height classes are chosen. First the directional bins are determined in a way that the sum of the energy fluxes of waves within each bin is equal. After determining the limits of directional bins, using the same principle of having equal energy flux the wave height classes in each directional bin is determined. Therefore the sum of energy fluxes in all cluster of waves are equal.

The representative wave height in this approach is calculated based on the average energy flux in each cluster using the following equation.

$$H_s = \sqrt{\frac{8.\overline{E_f}}{\rho.g.C_g}} \qquad (5)$$

Three scenarios are chosen for this approach. In first scenario the whole wave climate is schematized into nine representative wave conditions including three directional bins and three wave height classes, in the second scenario it is schematized in six wave conditions with three directional bins and two wave height classes. Also in the third scenario, the schematized wave climate includes six wave conditions but with six directional bins and one wave height class. Figure 6-6 shows a schematic plot for the wave schematization based on the energy flux for 3 directional bins and 3 wave height classes. All the representative wave conditions for this approach are summarized in Appendix.

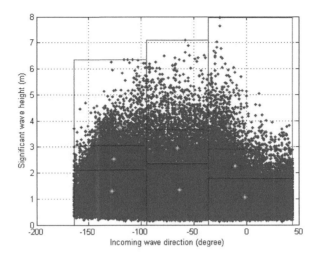

Figure 6 - 6 Representative wave conditions for scenario one of Energy Flux Approach (three directional bins and three wave height classes); border of the bins (red lines), all wave records (blue dots), and representative wave conditions (green dots)

" Opti " Approach

The wave climate can also be schematized using the so called optimal or 'Opti' approach. This approach aims to select an optimum subset of wave conditions that contribute more to the bottom change or residual sediment transport by elimination race algorithm. Initially starting with a broad range of wave conditions and carrying out a short morphological simulation for each wave condition, a map of erosion/deposition or residual sediment transport is produced for each wave condition. Then a target map is produced by computing a weighted average of these maps on the basis of the weighted sum of all of the wave conditions, each weighted by its probability of occurrence in the wave climate. A loop is then performed where wave classes are progressively omitted by dropping the condition which contributes the least to the resulting erosion/deposition or residual sediment transport. The weight (probability of occurrence) of the dropped condition is added to the most closely correlated remaining condition, and an inner loop is performed where random factors between zero and two are applied to the weightings of up to 3 of the most closely correlated remaining conditions and the RMS error between the new weighted sum of erosion/deposition or residual sediment transports, and the target result is computed. This loop is iterated many times and the randomly assigned class of weights which minimizes the RMS error is chosen. Again the condition which contributes least to the result is then dropped, and the outer loop repeats.

At each step the statistical parameters computed are as follows:

- rms, the root-mean-square error;
- stdc, the standard deviation of the approximating pattern;
- corr, the correlation coefficient between both patterns;

Based on these parameters and the acceptable rms, the modeler can choose the number of wave conditions for the simulation.

In this study we started the Opti procedure with 48 wave conditions (Table 6-1) and used the deposition/erosion maps as the criteria for the target solution. The statistical parameters computed by Opti method as a function of number of wave conditions is shown in Figure 6-7. The error increases with reducing the representative wave conditions, simply because there is no combination of remaining wave conditions that can produce the same pattern as the target map. Based on outcome of the algorithm we have chosen the last 9 remaining wave conditions to be as the schematized set of wave conditions.

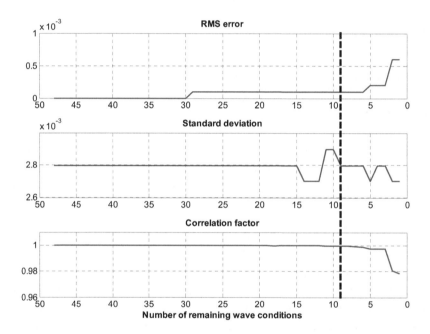

Figure 6 - 7 Statistical parameters of optimization as function of number of wave conditions

Summary of the representative nine wave conditions with the related weighted average obtained from 'Opti' is shown in Table 6-2.

Table 6 - 2 Representative wave conditions of 'Opti'

No. conditions	Dir. Range	Rep. dir. (^{0})	Rep. Hs (m)	Rep. Tp (s)	occ.%
1	-15-15	359	0.642	5.48	14.2
2	-15-15	357	1.399	6.42	32.7
3	195-225	213	0.716	4.49	1.1
4	195-225	217	2.404	6.46	11.0
5	225-255	239	0.671	4.70	2.9
6	255-285	270	2.418	6.62	10.7
7	285-315	301	0.622	5.27	15.0
8	285-315	301	1.452	6.05	9.3
9	285-315	300	4.390	8.62	3.2

Summary of Wave input Reduction approaches

In total, based on three different approaches, eight scenarios for wave schematization were developed (Table 6-3) These wave conditions are visualized in Figure 6-8 .

Table 6 - 3 Summary of all wave conditions applied in the model

	No. of wave height classes	No. of wave direction classes	Total wave conditions
Potential sediment transport approach			
1.	3	2	9
	1	3	
2.	2	3	6
3.	1	6	6
4.	3	2	9(each season)
	1	3	
Energy flux approach			
5.	3	3	9
6.	2	3	6
7.	1	6	6
'Opti' approach			
8.	-	-	9

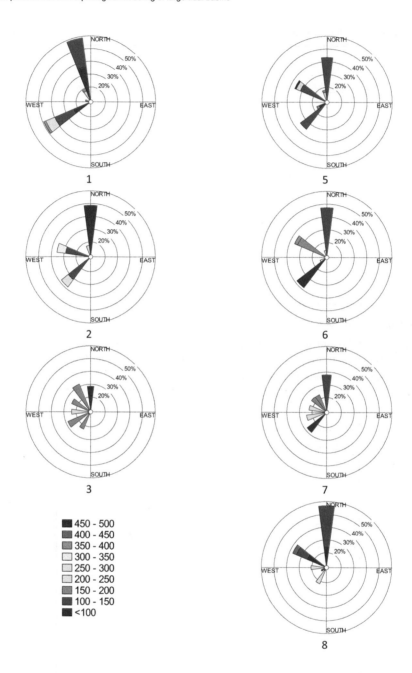

Figure 6 - 8 Visual representation of resulting wave conditions for different schematization approaches based on Table 6-3 (The seasonal schematization is not shown, wave height in cm)

6.4.3   Wind

In this study, based on the available data the wind speed is related to the significant wave height. This data-based relation is shown in Figure 6-9, therefore for each wave condition the wind speed can be determined based on this relation. The determined wind speed for each representative wave condition was also calculated using Brettschneider formula assuming un-limited fetch condition, and both methods gave similar values of wind speed for certain wave height. An example of the calculated wind speed for a specific wave height is shown in the Table 6-4. Also, it was assumed that the direction of the wind is the same as the related schematized wave direction.

Table 6 - 4 Wind speed calculated from Brettschneider and correlation obtained from the data

Wave height (m)	Wind speed (m/s)	
	Brettschneider	Correlation obtained from data, shown in Figure 6-9
0.67	4.83	4.99
1.43	7.04	7.54

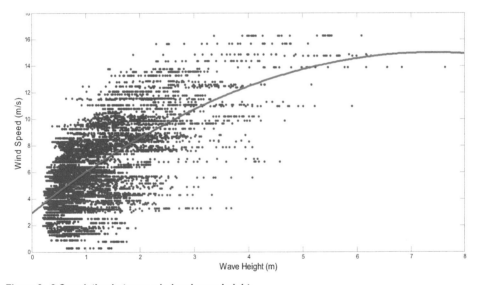

Figure 6 - 9 Correlation between wind and wave height

6.4.4 Surge

In a similar approach as with the wind schematization a data-based relation between wind speed, wind direction and surge (wind setup) at Texel inlet is generated. For the determined wind speed and direction in each wave condition the corresponding setup is calculated. This setup is added to the water level boundaries of the model for the duration of the respected wave condition.

6.4.5 Applying schematized wave conditions together with morphological factor

Regardless of which techniques are used to schematize the model inputs, the schematized forces (specifically wave conditions and morphological tide) must be turned into a morphological simulation of the required actual time. In order to combine different wave conditions with morphological tide while using a morphological factor, we have used two different techniques: Variable MorFac and Mor-Merge

Variable MorFac

Combining different wave conditions with morphological tide is extremely simple to achieve by changing the morphological factor during the simulation. In order to account for the random phasing between waves and tides that occurs in nature, each wave condition is simulated for the duration of one morphological tide of fixed (hydrodynamic) duration. A morphological acceleration factor specific to the wave condition is applied so that the morphological duration of the wave class matches its probability of occurrence. The MorFac required is computed by

$$MorFac = \frac{P_c \times T}{T_{mor.tide}}$$

In which:

P_c : probability of occurrence of the wave condition

T : Duration of the morphological simulation

$T_{mor.tide}$: Hydrodynamic duration of morphological tide

The models are set up in a way that a complete set of wave conditions is applied in each simulated morphological year, therefore for example in a 15 year long morphological simulation with 9 wave conditions a complete set of 9 wave conditions is repeated 15 times.

Special care needs to be taken when changing MorFac values between wave classes, especially when simultaneously changing the wave boundary condition, as it is important that approximately the same suspended sediment concentrations exist at the start and end of a MorFac value, otherwise the sediment mass balance will be altered. Therefore we have adopted a similar approach to Lesser (2009):

1. Set MorFac to zero at the end of a morphological tide (this stops the morphological simulation),
2. Change the required hydrodynamic model boundary conditions (wind, waves, surge, etc) over a period of 120 minutes to avoid shocking the models,
3. Allow the hydrodynamic and wave models to stabilize to the new boundary conditions over the rest of the current tidal cycle with MorFac of 0,
4. Set MorFac to the appropriate value for the new wave class, (this restarts the morphological simulation),
5. Compute hydrodynamics, sediment transport, and accelerated morphological change through exactly one morphological tide,
6. Set MorFac to zero to halt the morphological simulation,
7. Repeat the above steps until the end of the simulation.

The above steps ensure that the amount of suspended sediment is similar when changing the MorFac and the related wave conditions thus the discontinuity of mass is minimized. It is preferred to get almost the same average value of MorFac for different wave schematization approaches to avoid the influence of having very different MorFacs when analyzing and comparing different approaches. In the case where the MorFac value of a certain wave condition (with high probability of occurrence) is very high, which in turn makes the average MorFac of overall wave conditions deviate highly from the average in other schematization approaches, this MorFac value is lowered simply by allowing two or more morphological tides for the corresponding wave condition.

Mor-Merge

Mor-Merge is based on the Parallel Online Approach (Roelvink, 2006). The basis of this approach is the difference between the time-scale of hydrodynamic changes and morphology changes. The hydrodynamic conditions vary much faster than the morphology. If all the hydrodynamic conditions may occur during a small time interval compared to the morphological time-scale, they may as well occur simultaneously. Therefore if these conditions share the same bathymetry, they can be modeled in parallel in the morphological model. The bathymetry should be updated according to the weighted average of the changes in the bed level for all the conditions.

Using this approach the different wave conditions together with the morphological tide are simulated with the same constant MorFac on the same bathymetry and in parallel, then the resulting bed level change of all these simulations, are merged by weighted averaging, based on the probability of occurrence of the corresponding wave condition, to one new bathymetry for the next step. In this approach a series of PC's, a PC with multiple processors, a cluster, or a cloud may be used to model the parallel simulations.

6.5   Effect of Wave Schematization

In this section we have investigated the effect of different wave schematization methods on the morphological simulation of the Dutch Waddenzee by carrying out different hindcasting simulation for 15 years (1991-2005) using different sets of schematized wave conditions.

6.5.1 Model Setup

The main processes considered in the comparison are the tidal currents, waves, sediment transport and morphological change. Based on these processes the following model is set up.

Grid

The computational grid in this model is covering the area of interest, Marsdiep, Eierlandse Gat, Vlie and Ameland inlets. The same grid was used for both the Flow and the Wave model, but the grid of the wave was extended to the north east and to the south west to avoid disturbances at the boundary of the flow model. The main challenge when making the grid was to have grid cells large enough to result in reasonable computational time for the long study period (15 morphological years). On the other hand, the grid should be fine enough to show some details of necessary process especially in the surf zone where most of the wave energy dissipates due to wave breaking. Both grids are shown in Figure 6 -10.

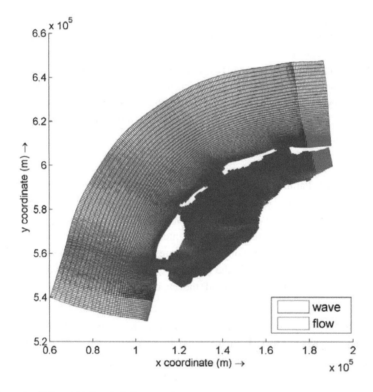

Figure 6 - 10 Computational gird

Bathymetry

The study period extended over 15 morphological years starting from the year of 1990. The bathymetry of 1990 based on the available data from The National Institute for Coastal and Marine Management (RIKZ) is used as the initial bathymetry of the model (Figure 6-11).

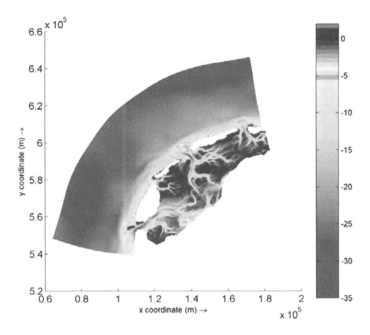

Figure 6 - 11 Bathymetry of the Waddenzee (1990)

Flow boundary condition

This model is forced by three boundaries; one water level boundary at the offshore boundary applied as a set of harmonic functions and two water level gradient boundaries at the lateral boundaries. In this model the only M2, M4 and M6 components are used to from the morphological tide (See section 6.4.1)

Wave boundary conditions

To apply the wave conditions in this model the variable MorFac method has been used and a separate simulation for each set of 8 set of schematized wave conditions is carried out. These simulations are listed in Table 6-5. Also in this table there are 2 additional simulations (9 and 10). These two simulations are carried out without waves. Therefore by comparing the result of the first 8 simulations with these two it is possible to distinguish the effect of the wave in the simulations.

Table 6 - 5 List of the simulations with different schematized waves.

No.	Simulation code	Simulation characteristics			
		Approach of schematization	No. of wave height classes (Hs)	No. of wave direction classes	Total wave conditions
1	SedP9	Potential sediment transport	3 / 1	2 / 3	9
2	SedP6a	Potential sediment transport	2	3	6
3	SedP6b	Potential sediment transport	1	6	6
4	SedP9s	Potential sediment transport - with seasons	3 / 1	2 / 3	9
5	Eflux9	Energy flux	3	3	9
6	Eflux6a	Energy flux	2	3	6
7	Eflux6b	Energy flux	1	6	6
8	Opti	'Opti'	-	-	9
9	TavgMf	Only tide with average morfac (Mf=78.4)	-	-	-
10	TvarMf	Only tide with variable morfac	-	-	-

Erosion of dry cells

One of the main problems facing the traditional numerical morphological models is the bank erosion. A simple approach is used in Delft3D Flow to solve the problem. This approach is called "dry cell erosion" as described by Roelvink et al. (2006). The parameter in Delft3D Flow; the factor for erosion of adjacent dry cells is set to 1.0 meaning that all erosion that would occur in the wet cell is assigned to the adjacent dry cells.

Sediment properties

In this model one fraction of non-cohesive sediment with D50 of 0.250 mm was applied to the model.

Sediment transport formula

The sediment transport relation used in the simulations of this study is based on the Soulsby - Van Rijn relation described by Soulsby (1997). This relation is a semi-empirical relation for the sediment transport in combined wave and current flow field.

6.5.2 Impact of waves

The influence of introducing waves in the model can be shown by comparing a simulation including both waves and tide with the one including only tides. In other words the simulation with tidal forcing is considered as the reference simulation. Simulations; *SedP6a* and *TvarMf* are compared because both of them have the same variable Morfac so the effect of applying different morphological factors does not affect the comparison.

Marsdiep basin and Texel inlet

Figure 6-12 shows the comparison between the resulting bathymetry from simulation with and without wave. It show that not only the wave action has a prominent effect on the ebb-tidal delta, but also the waves penetrated into the basin together with the waves generated by wind inside the basin affect the inter-tidal flats.

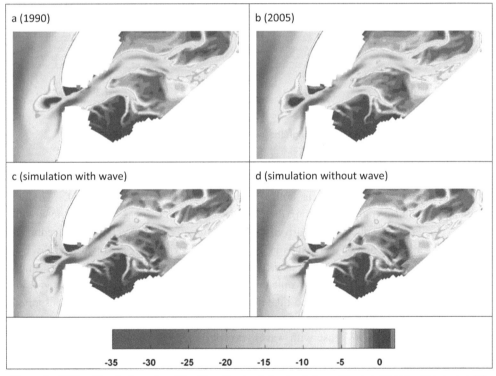

Figure 6 - 12 Resulting bathymetry of the Marsdiep basin in simulation with wave (c) and without wave (d) compared with the bathymetry of 1990 (a) and 2005 (b)

Sediment transport pattern

Figure 6-13 shows the net sediment transport through different inlets in both cases. The same pattern of sediment transport is still valid after implementing the waves, where the Texel inlet imports sediment while Eierlandse and Vlie export sediment. As a result of adding waves the amount of sediment transported through Texel increases by 5%. Also, it decreases in Eierlandse and Vlie by 29% and 47% respectively, which indicates that more sediment is transported towards the inlet due to the wave forcing. Implementing wave forcing has more influence on the sediment transport through Vlie inlet with the maximum relative difference compared to the simulation without wave.

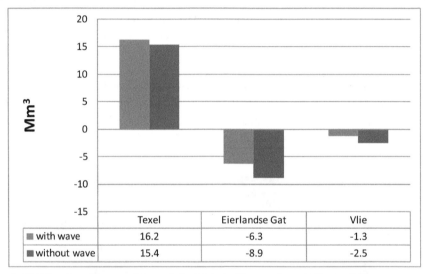

	Texel	Eierlandse Gat	Vlie
■ with wave	16.2	-6.3	-1.3
■ without wave	15.4	-8.9	-2.5

Figure 6 - 13 Net sediment transport through the inlets in 15 years. import (+), export (-)

Net sediment transport patterns for both simulations are shown in Figure 6-14. The net amount of sediment transported along the coast is much larger in case of simulations with waves. This is mainly due to wave driven longshore currents. The direction of the sediment transport at the shoreline of the islands in the simulations with waves is mainly northward, except at the southern tip of Texel Island because the sediment transport in that location might be influenced by the hydrodynamic pattern of the inlet and this area is protected from the winds from south-west by the ebb-tidal delta of the Texel inlet. The sediment transported along the shoreline of the islands in the simulation including only tide is very small and negligible comparing to the case with waves, except directly at the south of the Vlie inlet. There it is even larger than in the case with waves and it has a southward direction, due to the strong ebb-current at the Vlie inlet, when there is no wave forcing.

Figure 6 - 14 Total sediment transport (in Mm3) over 15 years; with waves (orange), without waves (white)

Interaction between tidal basins

The magnitude of sediment transported is varying spatially, thus the amount of transported sediment is different over the different sections of a tidal divide depending on the location, but our concern is the direction of net sediment transport over each tidal divide. The direction of net sediment transport has not changed when the wave forcing was applied (Figure 6-15), but the amount of net sediment transport has changed. This difference is very small except for the net sediment transport from Eierlandse to Marsdiep which got much larger (4 times more) with the waves. In spite of such increase, this amount is still considerably small.

After calculating the net sediment transport through the inlets and over tidal divides for both simulations, the sediment change in each basin can be calculated. Figure 6-16 presents the amount of sediment change in each basin. Positive values mean that the basin is gaining sediment while negative values means that the basin is losing sediment. The Marsdiep tidal basin imports a substantial amount of sediment through the Texel inlet and a small amount from Eierlandse Gat basin. Part of the imported sediment is exported to the adjacent Vlie inlet. In the case of the simulations with waves the Marsdiep basin gains 11.6% more sediment comparing to the simulation without waves. Eierlandse Gat basin exports more sediment than what it imports, thus it loses sediment, but the amount of sediment lost decreased by about 25% when the wave forcing is included.

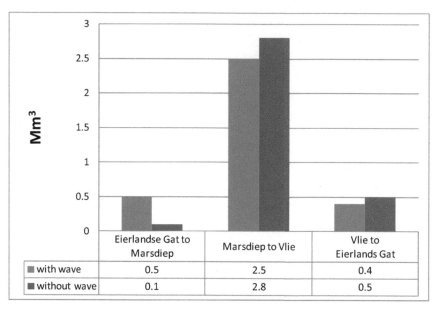

Figure 6 - 15 Net sediment transport between the basins in 15 years (over tidal divides)

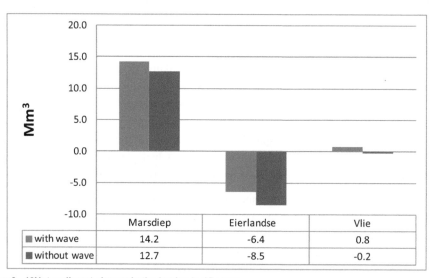

Figure 6 - 16Net sediment change in the basins in 15 years

6.5.3 Comparison between approaches of wave schematization

In this section the difference between different approaches of wave schematization is investigated. The comparison is focused on the sediment transport, change of sediment volume each tidal basin and changes of the ebb tidal delta of Texel inlet.

Sediment transport

Figure 6-17 shows the sediment transport through the inlets for different wave schematization approaches. The sediment transport through the Eierlandse Gat inlet in different simulations is almost the same while there is a small change in both Texel and Vlie inlets. Looking closer to the figure, it can be noticed that the amount of sediment transport through Texel and Vlie follows a trend: In Texel inlet, the sediment import through the inlet is more for the wave schematization with more directional bins and less wave height classes. At the Vlie inlet the export of sediment follows the same trend.

The sediment transport through Texel inlet due to wave schematization using 'Opti' method is similar to that resulting from wave schematization of more directional bins in potential sediment transport approach and the energy flux approaches. However, in the Vlie inlet it is similar to the approaches with more wave height classes.

The sediment transports between the basins over the tidal divides for different approaches of wave schematization are presented in Figure 6-18. There is no considerable difference between the results of different approaches. Despite the small differences, it can be noticed that the amount of sediment transport for the scenarios having the same number of wave height classes and directional bins regardless of the approach of the schematization is almost the same. And the result of the *Opti* method is similar to schematization with less directional bins.

Since the transport through the tidal divides is similar and the very small differences are negligible comparing to the total sediment change in the basins, the total sediment change in the basins is highly determined by the transport through the inlets. That is why total sediment change in the basins reveals a similar trend as the amount of sediment transport through the inlets (Figure 6-19).

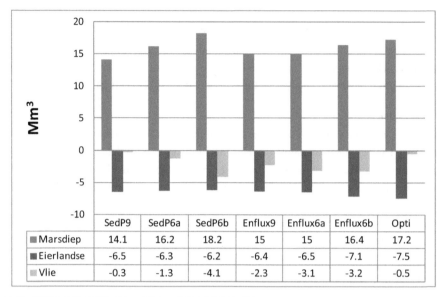

Figure 6 - 17 Net sediment transport through the inlets in 15 years

	SedP9	SedP6a	SedP6b	Enflux9	Enflux6a	Enflux6b	Opti
■ Marsdiep	14.1	16.2	18.2	15	15	16.4	17.2
■ Eierlandse	-6.5	-6.3	-6.2	-6.4	-6.5	-7.1	-7.5
▪ Vlie	-0.3	-1.3	-4.1	-2.3	-3.1	-3.2	-0.5

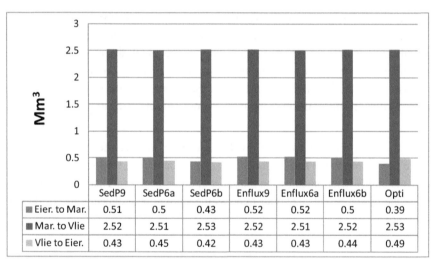

Figure 6 - 18 Net sediment transport between basins in 15 years

	SedP9	SedP6a	SedP6b	Enflux9	Enflux6a	Enflux6b	Opti
■ Eier. to Mar.	0.51	0.5	0.43	0.52	0.52	0.5	0.39
■ Mar. to Vlie	2.52	2.51	2.53	2.52	2.51	2.52	2.53
▪ Vlie to Eier.	0.43	0.45	0.42	0.43	0.43	0.44	0.49

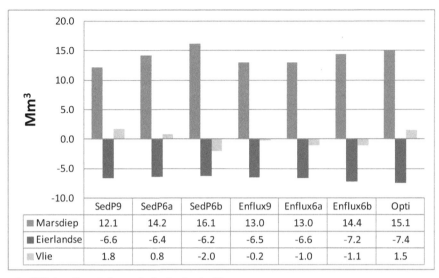

	SedP9	SedP6a	SedP6b	Enflux9	Enflux6a	Enflux6b	Opti
■ Marsdiep	12.1	14.2	16.1	13.0	13.0	14.4	15.1
■ Eierlandse	-6.6	-6.4	-6.2	-6.5	-6.6	-7.2	-7.4
▨ Vlie	1.8	0.8	-2.0	-0.2	-1.0	-1.1	1.5

Figure 6 - 19 Net sediment change in the basins in 15 years

Ebb tidal delta

The change in the volume of the Texel ebb tidal delta due to applying different wave schematizations is presented in Figure 6-20. For the potential sediment transport approach, changes in volume for *SedP9* and *SedP6a* simulations show a similar pattern. In these two simulations, the volume of the Texel delta increased slightly during the first five years, then started to decrease with higher rate. The trend in the simulation SedPg6b is different as Texel inlet gained sediment till the year of 2000 then started to lose sediment and the final volume is almost same as the initial one.

For the Energy flux schematization approach, the *Eflux9* and *Eflux6a* simulations resulted in a similar change of Texel ebb tidal volume. The volume of the delta increased at a low rate during the first ten years, then decreased at higher rate. The volume at the end of the two simulations is 98% of the initial volume. *Eflux6b* shows a different pattern as the volume increased substantially during the first year then started to lose sediment. Still the final volume is greater than the initial.

Therefore, it can be concluded that amongst the scenarios implemented in the model, the schematization having more wave height classes and less directional bin causes more erosion in the ebb tidal delta. Also overall, Energy flux schematizations results in larger volume of the Texel ebb tidal delta over the simulation period comparing to potential sediment transport approach. The maximum erosion of the Texel ebb tidal delta occurred with the wave schematization using the 'Opti' approach. Despite the fact that the wave rose of schematized wave conditions (Figure 6-8) visually has the lease similarity with the measured wave rose (Figure 6-2), the rate of volume reduction in Texel ebb-tidal delta is more similar to the rate computed based on bathymetry measurement data.

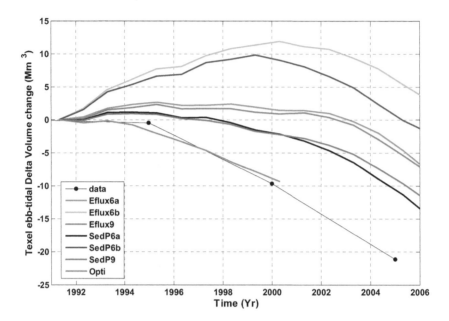

Figure 6 - 20 Change in volume of Texel ebb-tidal delta for different wave schematization approaches (filtered for changes within a year)

Effects of applying different seasons

The same sediment transport values as in the previous section are also compared for the simulations with and without the effect of seasonality (See section 6.4.2). These comparisons are shown in Figures 6-21 to 6-23. In the simulation with the effect of seasonality the sediment exchange at the inlets is more than the simulation without this effect. Also the volume change when the ebb-tidal delta of the Texel inlet is investigated for the effect of seasonality in schematizing wave forcing (Figure 6-24). Including seasonality in the wave schematization has a very clear effect on the inter-annual changes, however as it is clear in the picture, the differences are damped out every two years.

A note on the calculating volume of Texel ebb-tidal delta from measured bathymetry

In this study the bathymetric map for each year is used to calculate the volume of Texel ebb-tidal delta on that specific year, however the accuracy of the data in different years is not the same. The older the data the less accurate they are. But in this calculation the data for different years is assumed to be equally valid. The bathymetry of each year is interpolated on the same computational gird as the morphological model and the volume of ebb-tidal delta is calculated based on Walton and Adams (1976) method. The cross-shore profile of the beach at the Texel Island out of the influence of ebb-tidal delta is considered as undisturbed coastline in the Walton and Adams (1976) method.

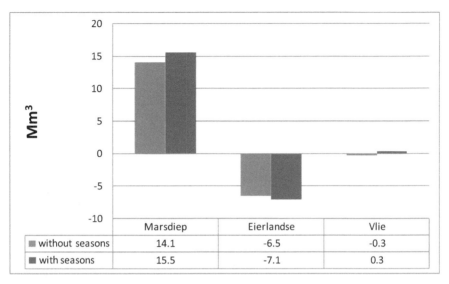

	Marsdiep	Eierlandse	Vlie
▪ without seasons	14.1	-6.5	-0.3
▪ with seasons	15.5	-7.1	0.3

Figure 6 - 21 Net sediment transport through the inlets in 15 years

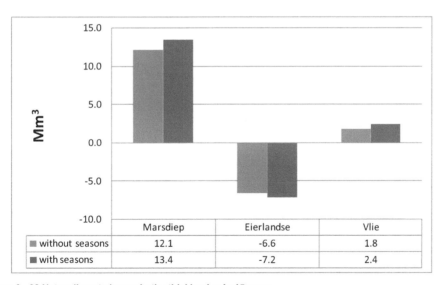

	Marsdiep	Eierlandse	Vlie
▪ without seasons	12.1	-6.6	1.8
▪ with seasons	13.4	-7.2	2.4

Figure 6 - 22 Net sediment change in the tidal basins in 15 years

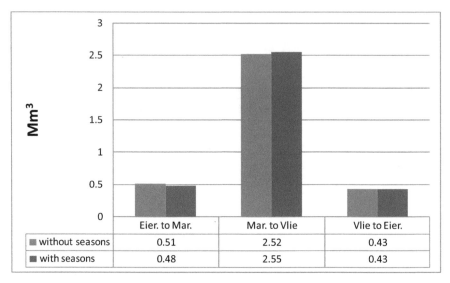

	Eier. to Mar.	Mar. to Vlie	Vlie to Eier.
■ without seasons	0.51	2.52	0.43
■ with seasons	0.48	2.55	0.43

Figure 6 - 23 Net sediment transport between tidal basins over the tidal divides in 15 years

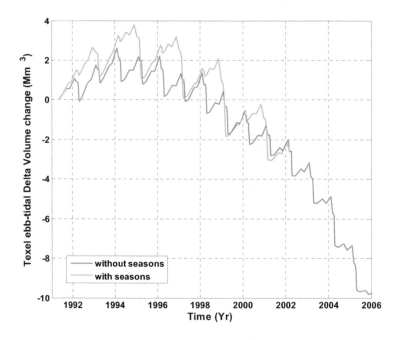

Figure 6 - 24 Change in volume of Texel ebb-tidal delta for the effect of the applying seasonality in the wave schematization

6.5.4   Effect of using variable MorFac comparing to a single MorFac

In section 6.5.2, it was aimed to isolate the effect of waves by comparing two simulations one with both wave and tidal forcing and one with only tide. However, to make them comparable the simulation without wave was also carried out with the same variable MorFac as the simulation with wave. But there is a need to pinpoint the effect of applying variable MorFac in the simulations, therefore in this section we compared two simulations; the first is *TvarMf* which includes variable MorFac and the second is *TavgMf* which has one MorFac value, this value is the average of MorFac values in the first simulation. Tables 6 to 8 show that at the time scale of this type of studies there is no remarkable difference between a simulation with variable MorFac and a simulation with constant MorFac with the morphological factor of the same order.

Table 6 - 6 Net volume of sediment transported through the inlets, (+) towards the basin, (-) outside the basin

Inlet	Vol. of sed. transport (Mm^3)	
	(*TavgMf*)	(*TvarMf*)
Texel	15.6	15.4
Eierlandse Gat	-8.2	-8.9
Vlie	-0.2	-0.2

Table 6 - 7 Net sediment transported between the basins in 15 years (over tidal divides)

Direction of net sediment transported	Vol. of sed. transport (Mm^3)	
	(*TavgMf*)	(*TvarMf*)
Eierlandse Gat to Marsdiep	0.11	0.11
Marsdiep to Vlie	2.87	2.87
Vlie to Eierlandse Gat	0.57	0.55

Table 6 - 8 Net sediment change in the basins in 15 years

basin	Vol. of sed. (Mm^3)	
	(*TavgMf*)	(*TvarMf*)
Marsdiep	12.9	12.6
Eierlandse Gat	-7.7	-8.5
Vlie	1.8	1.8

6.6   Hindcasting

The morphological evolution of the Dutch Waddenzee since the closure of the closure of the Zuiderzee can be divided roughly into two different periods. The first period, roughly between 1930 and 1980, is the period in which the morphological changes are mainly driven by the effect of the construction of the closure dam and these effects dominate the natural evolution. In the second period starting around 1980, the effect of the closure of the basin is damped to a large extent and the morphological evolution follows more natural course. In this section we have tried to use the techniques and knowledge developed in this dissertation to hindcast the first period of this morphological evolution.

6.6.1   Model setup

Grid

Figure 6-25 shows the wave grid and flow grid used in the models. The grid cells have an average resolution of about 298m x 509m but are refined with a resolution of 34m x 66m in and around the Texel-Marsdiep channel.

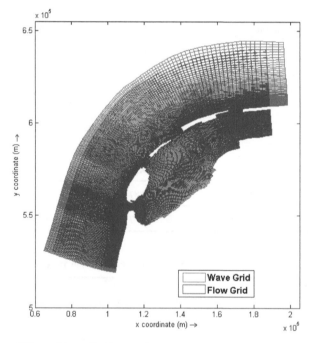

Figure 6 - 25 Flow and Wave grids applied in models

Bathymetry

For the hindcast simulations, the bathymetry of 1930 (Figure 6-26) from the National Institute for Coastal and Marine Management (RIKZ) is used as the initial bathymetry of the simulations.

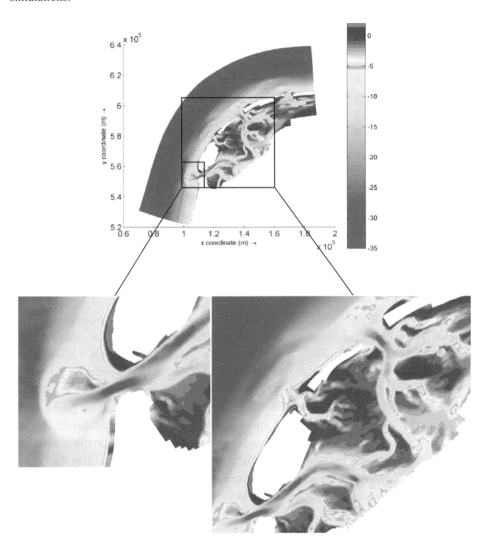

Figure 6 - 26 Initial bathymetry of the hindcasting simulations (year 1930) with close view of the ebb-tidal delta of the Texel and the basins of the Western Dutch Waddenzee

Flow boundary condition

This model is forced along three boundaries; one water level boundary at the offshore boundary applied as a set of harmonic functions and two water level gradient boundaries at the lateral boundaries. In this model the only M2, M4, M6, and the artificial tidal constituent. C1 components are used to from the morphological tide (Section 6.4.1).

Wave Boundary Condition

To apply the wave conditions in this model both the variable MorFac and the Mor-Merge method have been used and a separate simulation for each method is carried out. The schematized wave conditions including 6 wave conditions based on wave energy flux approach are chosen for the hindcast simulations, with 3 directorial bins and two wave height classes.

Sediment Properties

As it was shown in chapter 4, if only one single sediment size is used in the model, mechanisms such as the armoring of the channel bed due to the erosion of finer materials and sediment sorting are systematically neglected in long-term morphological simulations. Therefore for the simulations a spatial distribution of eight sediment fractions is used based on the methodologies introduced in chapter 4.

6.7   Results and Discussions

Figure 6-27 shows the measured bathymetry of the Texel ebb-tidal delta (Noordehaaks) and the adjacent channels in comparison with the result of both approaches of the simulations, as it is clear in the picture the ebb-tidal delta is not reproduced well enough. Especially the model is not able to reproduce the dry part of the ebb-tidal delta. In the model results only in the case of Mor-Merge in the year of 1940 and 1950 there are some very small areas with the elevation of more than 0. These areas are also eroded later during the remaining time of the simulation. This difference between the simulations is mainly due to the use of dry cell erosion technique in the simulations. However if this technique is not used in the simulations, the migration of the shoals inside the basin will be much limited. Therefore using the dry cell erosion technique should be based on the features which are the main interest of different simulations. The behavior of the channels however is reproduced better in the simulations, especially with the Mor-Merge approach. The main ebb-channel in front of the inlet is bending toward the south and a smaller channel is produced at the north side of the ebb-tidal delta. Overall it seems that the model shows a more diffusive sediment transport comparing to the reality. The developing shoal at the southern tip of the Texel Island (Hors) is simulated better in the model with the Variable MorFac approach. Both simulations are showing more or less the same end results but in different time scales : the Variable MorFac approach seems to reach some degree of stability in the year 1960, with a resulting bathymetry similar to the resulting bathymetry of the Mor-Merge approach but in the year 1970. This suggests that the Variable MorFac in this case has a higher speed of morphological changes than the Mor-

Merge approach. Looking at both Figures 6-27 and 6-28, it is obvious that the resulting bathymetry of the Variable MorFac approach produces more sediment deposition on the shoal areas especially in the Eierlandse Gat basin. This over estimation of sediment deposits can be due to the time of changing the MorFac in Variable MorFac approach. In this approach the MorFac should change at the time with the lowest concentration of suspended sediment, however due to the propagation of tidal wave in and outside of the Dutch Waddenzee tidal basins, the minimum concentration of suspended sediment in different tidal basins does not coincide. In this simulation the condition of lowest concentration of suspended sediment is fulfilled for the Marsdiep basin which does not necessary mean that it is the same in Eierlandse Gat basin. On the other hand the Mor-Merge approach produces shallower shoals than the measured data. This smaller deposition can be caused by over-estimation of the wave action inside the basins. The same concept is shown in the resulting erosion sedimentation patterns as well (Figure 6-29).

Figure 6-30 shows the change in volume of Texel ebb-tidal delta for different modeling approaches and the measured data. It shows that in the first 5 years in both methods the volume of ebb-tidal delta is reducing unrealistically, but in the Mor-Merge this volume stabilizes and reaches the same value of measured data. Also figure 6-31 shows the changes of the skill of the simulations with different approaches based on the Brier skill score. Despite the fact that based on this skill score all the results of the simulations are considered bad, interestingly this figure demonstrates that in the simulation with the Mor-Merge approach the skill of the simulation increases with time and finally gets to positive values at the year of 1980. The increase of the skill score does not appear in the simulation with the Variable MorFac, which may be due to the higher speed of morphological changes in this approach. Therefore from these two figures we can conclude that although a lot of effort has been done to reduce the spin-up time of the model, the morphological simulation during the early times of the run is adjusting the bathymetry to the numerical parameters of its equations rather than the natural physical processes.

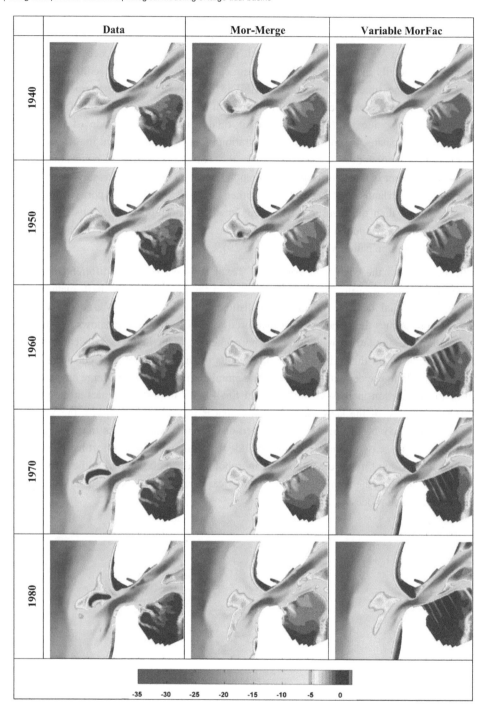

Figure 6 - 27 Comparing the measured bathymetry of the ebb-tidal delta of the Texel inlet with the model results

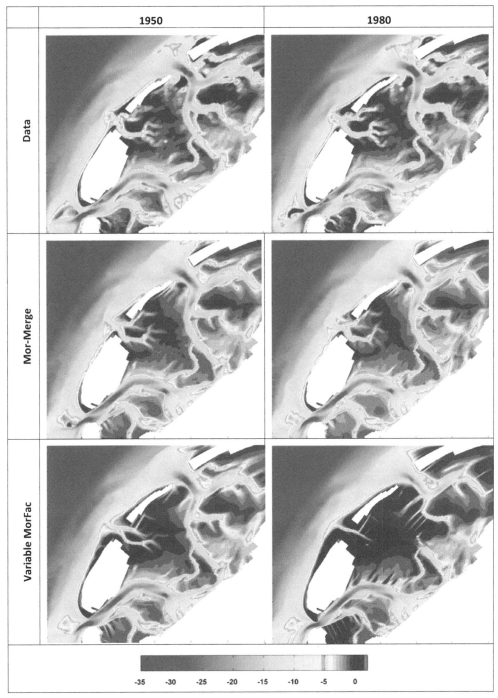

Figure 6 - 28 Comparing the measured bathymetry of the Western Dutch Waddenzee with the model results

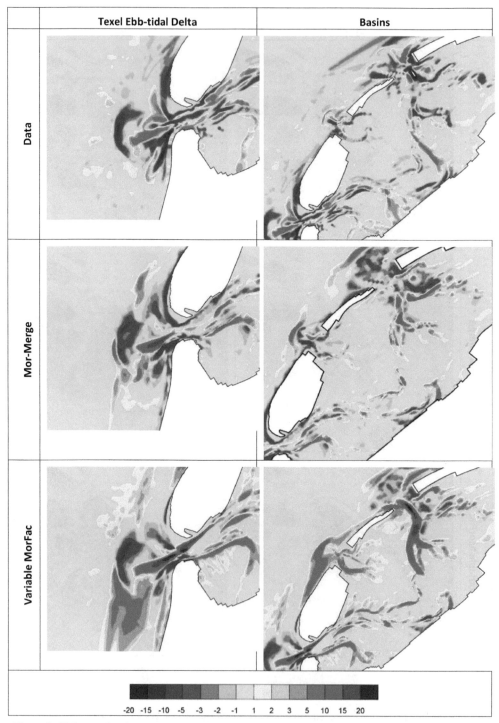

Figure 6 - 29 Comparing the measured erosion deposition patterns with the model results

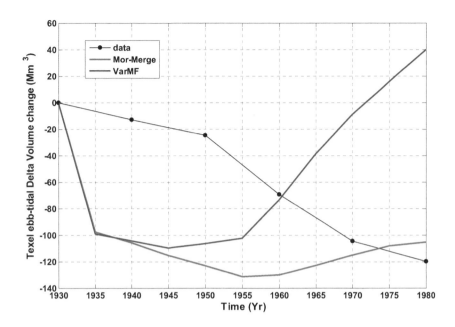

Figure 6 - 30 Change in volume of Texel ebb-tidal delta for different modeling approaches (filtered for changes within a year)

A note on the computational time in Mor-Merge and Variable MorFac approaches.

The main difference between Mor-Merge and Variable MorFac approaches in the sense of computational efficiency is the fact that the simulations with the Variable MorFac approach can be carried out on a single processor PC but the simulations with Mor-Merge approach need more computational nodes (No. of simulations for different forcing conditions + 1). However in the case that there is a limited number of computational nodes available for the simulation, a simulation with Variable MorFac approach can make use of all the nodes simultaneously, but in a simulations with Mor-Merge approach each simulation will use one (or one cluster of) node(s) for the simulation of each forcing condition and these simulations should wait for each other at each time step to be able to merge the bathymetry correctly. Therefore the Variable MorFac approach uses the computational power more efficiently. On the other hand, since we need to have tidal cycles with MorFac of zero, if the average MorFac of Variable MorFacs is equal to the one used in Mor-Merge, for the same morphological time the Variable MorFac approach needs two times more hydrodynamic time.

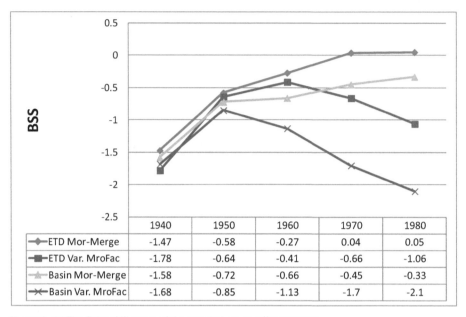

Figure 6 - 31 The Brier skill score of the simulations at different years

6.8 Conclusions

The conclusion of this chapter can be divided in three parts : The first part includes the influence of adding waves to the simulations with only tidal forcing, the second part is related to the effect of different wave schematization approaches, and the third part deals with the hindcast simulation and the approaches used for that purpose:

Adding wave forcing has a substantial effect on the entire tidal inlet system and the main changes can be represented by :

- The ebb tidal delta experienced big changes compared to the simulation with only tide: Volume of Texel ebb-tidal delta decreased to about 65% as a result of implementing waves;

- Longshore sediment transport along the outer coast of the barrier islands increased to a large extent;

- The waves push more sediment towards the inlets. Therefore Texel inlet imported more sediment with the waves while both the Eierlandse Gat and Vlie inlets exported less sediment. The change is quite obvious in Vlie inlet where the difference reached about 48%;

- As the result of waves Texel and Vlie basins gained more sediment, either through the inlets or from the tidal divides;

- Also given the size of the basins in the Dutch Waddenzee the waves generated by wind inside the basin have an important effect on the behavior of the shoals inside the basins.

The main findings from the different wave schematization approaches are:

- Though the difference in sediment transport through the inlet is relatively small, a trend can be noticed. The sediment transport through Texel inlet is higher for the wave schematization with more wave height classes and fewer directional bins. On the other hand, sediment transport through Vlie inlet is more for the schematized waves with more directional bins and less wave height classes. However, the result of applying schematized waves using the 'Opti' method is similar to schematization with more directional bins in Texel inlet and similar to schematization with more wave height classes in Vlie inlet.

- The total sediment changes in the Marsdiep and Vlie basins follow the same trend as the transport through their inlets.

- The resulting volume of Texel ebb-tidal delta is larger when implementing wave conditions with same directional bins and the wave height classes in Sediment transport potential and Energy flux approaches resulted in similar pattern of volume change. However waves schematized by the 'Opti' method caused the maximum erosion in Texel ebb-tidal delta.

- In addition, taking two seasons of summer and winter in the wave schematization did not cause a meaningful difference in the result of simulations on the time scale of the simulations.

The outcome of the hindcasting attempt can be summarized as :

- The current simulation with the schematization methods and modeling approaches are performing better in reproducing the behavior of the channels rather than that of the shoals, however the skill score of the simulation both inside the basins and on the ebb-tidal delta is considered bad.

- The simulated morphological evolution in the Variable MorFac approach is faster than in the Mor-Merge approach as well as the natural evolution.

- The result of the simulation with the Mor-Merge approach improves with time, suggesting that the model initially is adjusting to the numerical parameters of its equations rather than the physical processes.

- Although both hydrodynamic and morphological spin-up are considered in the simulation the model still spends some spin-up time overruling the natural physical processes by numerical processes.

CHAPTER 7

Conclusions and recommendations

This study is a part of a collective effort to bridge the existing gap of our understanding of morphological behavior of tidal basins between engineering and geological time scales by extending the use of coastal engineering tools (process-based models) to geological time scales (Figure 1-1). The hypothesis that 'If you put enough of the essential physics into the model, the most important features of the morphological behavior will come out, even at longer time scales' (Roelvink, 1999) is examined. This study shows that this hypothesis is valid and makes the relation between 'most important features ' and ' the essential physics ' more clear.

7.1   Response to research questions

To what extent does long-term process-based morphological modeling produce sensible results and which morphological features can be simulated by them?

In this study it is shown that a process-based model can be used to simulate long-term morphological changes in tidal basins and produce reasonable results. In chapter 2, the result of a very simplified model of the Dutch Waddenzee shows a good qualitative agreement with current pattern of channels and shoals of the Dutch Waddenzee, and also the morphological features of the basins such as area, volume and height of the inter-tidal flats follow the data-based equilibrium equations. However all of the simulations do not result in one single mega-scale stable (equilibrium) for all initial conditions, but with each initial condition in many aspects such as sediment exchange and some basic characteristics of tidal basins, a mega-scale stable (equilibrium) condition is simulated, which depends not only on the forcing boundary conditions but also on the initial condition. In chapter 2, we have also shown that such a simple process-based morphological model can reproduce some observed long-term morphological behavior of the Dutch Waddenzee mainly in the case of the effect of adjacent basins on each other and movement of tidal divides (tidal basin boundaries). In chapter 3, we demonstrated that the process-based morphological model is capable to qualitatively assess the long-term impacts of large scale human intervention in a coastal system, such as reproducing the change in tidal transport regime and the ensuing changes in morphological characteristics of the tidal basins due to extreme changes like the closure of the Zuiderzee in the case of the Dutch Waddenzee. In chapter 4 we compared the outcome of long-term morphodynamic simulations with conceptual models in the case of a medium size tidal basin with different wave and tide conditions. A good agreement between the morphological simulations and conceptual models on a decadal time scale is observed. However although adding processes to simulations and make them more complex helps the hindcasting simulations to perform better, hindcast simulation in chapter 3 and 6 did not reach high quantitative skill scores. These simulations were able to reproduce the behavior of main morphological features qualitatively well. Therefore we can conclude that this type of simulations at their present stage and our current knowledge is more helpful when used as realistic analogue rather than producing virtual reality and using them as realistic analogue can lead us to a better prediction of the future morphological state of the system in question.

What are the most important processes in long-term morphological modeling of tidal basins?

This research starts with a very simple simulation with only tidal forcing, geometry, and the simplest form of sediment transport. This simulation produced qualitatively good results, but also it revealed the need of more processes to obtain more realistic results (Chapter 2) . The first step was to add the sea bed sediment composition and distribution to the model clearly it showed that when we are dealing with long-term morphological simulations the sediment composition plays a significant role. Introducing a logical initial sediment size distribution improves the performance of the model significantly. However we are confronted with very limited available data in most of the cases. So, we suggested a few methodologies to generate this initial sediment size distribution by simply adapting the sediment distribution to the hydrodynamic condition at the beginning of the morphological simulation and introduced the concept of 'morphological spin-up' (Chapter 4). The next logical step was adding wave action to the model (Chapters 5 and 6). It is shown that not only the wave is an important process outside the inlet and at the location of the ebb-tidal delta, but also in the large tidal basins like the Dutch Waddenzee, the wind generated waves inside the basins are as important. In this study we introduced and examine different methodologies to schematize the wave climate for a long-term morphological simulation and showed that although the waves are very important, at the larger scale the chronology and the wave schematization approach are of less importance (Chapter 6)

What are the mechanisms governing the large scale morphological changes in tidal basins?

A morphological mechanism is defined as a (non-linear) interaction of different processes causing changes in a morphological feature. This study (chapter 2) shows that the mechanisms of generation of channel and shoal patterns and ebb-tidal deltas are mainly governed by the interaction of tidal forces with available sediment and geometrical boundary and an intervention can affect this mechanism through one of the processes and change or even reverse the trend of this mechanism. For example, closure dams can change the flow field due to tidal movement in a way that an exporting system become an importing system and a growing ebb-tidal delta starts shrinking in volume and in this scale the changes can be explained by only this mechanism (chapter 3). But some other more detailed features such as the maximum depth of channels and channel migration in the basins are governed by armoring of the channel bed and sediment sorting (chapter 4). Outside the basins the mechanisms which develop and re-shape the ebb-tidal delta and ebb and flood channels such as sediment by-passing and breaching and rotating of the channels includes the wave action as well. However the variation of wave directions is not an essential ingredient of these mechanisms but the relative importance of wave and tidal forces is a determining parameter (chapter 5).

Can long-term process-base models be used to assess the effect of human interventions on the evolution of tidal basins?

In this research (Chapter 3) a long-term process-based modeling approach is used to study the effect of one of the largest human intervention in tidal basins, The closure of the Zuiderzee, on the morphology of the adjacent tidal basin. The hydro- and morphodynamics for a total of five different scenarios were studied: a base scenario where no closure takes place, and four scenarios corresponding to different times of closure. The study shows that the model is capable to reproduce the change in tidal transport regime and proves that long-term process-based modeling is capable to qualitatively assess the long-term impacts of large scale human intervention in a coastal system.

7.2   Recommendations

7.2.1   Input schematization and climate change

In this study all the input schematization are carried out based on the historical measured data, without including any effect climate change. Arguably the climate change influences the key processes in the morphological evolution of tidal basins including but not limited to wave-wind climate, extreme storms, storm surge, sea level rise etc. (Nicholls, 2007). These influences should be accounted for in the input schematization for process-based simulations especially if they are used for forecasting (virtual reality). Although there have been some global and local studies to identify the influences of climate change on these processes, the predictions of the changes are significantly uncertain. Therefore a logical method for examining these effects is to develop some scenarios of these influences for the area of interest and include those scenarios in the input schematization (Bruneau et al. 2011, Dissayanake et al. 2012, Doung et al. 2012).

7.2.2   Sensitivity analyses for virtual reality simulations

A process-based morphological model, like the one which is used in this study, includes numerous numerical and physical parameters, most of these parameters are traditionally set for short term simulation and the available sensitivity analyses on these parameters are also limited to short term simulations. However a long-term simulation may be more sensitive to some of these parameters and default values of those parameters may not be suitable for longer simulations. Therefore there is a need for a systematic sensitivity analyses for long-term simulation which needs a considerable amount of computational time. To make the sensitivity analyses feasible a tidal inlet with small scale, such as the scale of the tidal basin in Chapter 5 of this study, can be used for sensitivity analyses in a hindcasting exercise. In the Dutch Waddenzee the Ameland inlet can be a good option.

7.2.3   Three dimensional processes

In this study all the simulations are carried out in the depth averaged fashion, therefore the three dimensional processes are neglected and the possible effect of these processes in the long-term morphological developments is not considered. In the Dutch Waddenzee the fresh water discharge to the basins and the salinity and concentration differences between inside the basins and the North Sea can cause stratification. Elias and Stive (2005) demonstrated the effect of stratification on the residual flow at the Texel Inlet based on some ADCP measurements. They showed a typical density-driven distribution of ebb-dominant flow in the top-layers and flood-dominant flow in the near-bed regions. The main challenge is how to evaluate the long-term effect of this type of processes and include them in a long-term simulation.

7.2.4   Fine and cohesive sediment

Since the percentage of fine and cohesive sediment is very low in the western Dutch Waddenzee, This type of sediment is not included in the simulations of the current study. However especially at the inter-tidal areas in which the mud content is larger than 20% this type of sediment can influence the morphological development, therefore a future study on the effect of this type of sediment on the long-term morphological development is necessary.

7.2.5   Evaluation of morphological simulations

As process-based models are able to simulate long-term morphological evolutions, there is a need for methods and tools to evaluate the performance of the simulations against observed data. In this research, conceptual models, empirical equilibrium relationships, volumes, hypsometry and visual patterns (qualitative method) together with Brier Skill Score (quantitative method) are used to evaluate the simulations. But each of these parameters has its own limitations, on one hand qualitative methods lack some degree of objectivity and on the other hand quantitative methods may easily neglect skill of a simulation in reproducing expected morphological features but in slightly different location. Therefore better pattern recognition algorithms that are able to classify and compare model results with observations, can help to evaluate the morphological simulations more comprehensively.

APPENDIX

Schematized wave conditions for different approaches

Table A - 1 Clusters of wave conditions selected for scenario 1 of Sediment transport potential approach (three dir. bins and three wave height classes)

%	N 345-15	NNE 15-45	ENE 45-75	E 75-105	ESE 105-135	SSE 135-165	S 165-195	SSW 195-225	WSW 225-255	W 255-285	WNW 285-315	NNW 315-345
0-1	7.19	4.87	2.26	1.02	0.74	0.83	1.17	3.33	4.81	3.69	4.38	6.33
1-2	6.09	2.78	1.91	0.75	0.43	0.35	0.81	4.79	5.80	4.06	4.52	7.12
2-3	1.34	0.51	0.38	0.07	0.01	0.03	0.11	1.89	2.89	2.04	1.93	2.88
3-4	0.24	0.07	0.01	0.00	0.01	0.00	0.02	0.39	0.99	0.78	0.79	0.97
4-5	0.04	0.00	0.00	0.00	0.00	0.00	0.00	0.03	0.18	0.31	0.29	0.31
5-6	0.01	0.00	0.00	0.00	0.00	0.00	0.00	0.00	0.02	0.10	0.07	0.08
6-7	0.00	0.00	0.00	0.00	0.00	0.00	0.00	0.00	0.01	0.02	0.02	0.02
7-8	0.00	0.00	0.00	0.00	0.00	0.00	0.00	0.00	0.00	0.00	0.00	0.00

Table A - 2 Wave climate schematization for scenario 1 of Sediment transport potential approach

No. conditions	Dir. Range	Rep. dir. (^{0})	Rep. Hs (m)	Rep. Tp (s)	occ.%	Wind spd. (m/s)	Wind dir. (^{0})	Morfac
1	195-285	241	0.67	4.71	13.3	4.99	241	93.8
2	285-45	346	0.64	5.41	25.6	4.86	346	180.7
3	195-285	239	1.46	5.60	16.4	7.64	239	116.1
4	285-45	340	1.43	6.31	23.0	7.54	340	162.6
5	195-285	241	2.42	6.55	7.6	10.55	241	54.1
6	285-45	332	2.40	7.13	7.4	10.51	332	52.8
7	195-255	234	3.59	7.46	1.8	13.59	234	12.9
8	255-315	284	3.95	7.95	2.6	14.44	284	18.9
9	315-45	337	3.85	8.64	1.9	14.21	337	13.4

Table A - 3 Clusters of wave conditions selected for scenario 2 of Sediment transport potential approach (three dir. Bins and two wave height classes)

%	N 345-15	NNE 15-45	ENE 45-75	E 75-105	ESE 105-135	SSE 135-165	S 165-195	SSW 195-225	WSW 225-255	W 255-285	WNW 285-315	NNW 315-345
0-1	7.19	4.87	2.26	1.02	0.74	0.83	1.17	3.33	4.81	3.69	4.38	6.33
1-2	6.09	2.78	1.91	0.75	0.43	0.35	0.81	4.79	5.80	4.06	4.52	7.12
2-3	1.34	0.51	0.38	0.07	0.01	0.03	0.11	1.89	2.89	2.04	1.93	2.88
3-4	0.24	0.07	0.01	0.00	0.01	0.00	0.02	0.39	0.99	0.78	0.79	0.97
4-5	0.04	0.00	0.00	0.00	0.00	0.00	0.00	0.03	0.18	0.31	0.29	0.31
5-6	0.01	0.00	0.00	0.00	0.00	0.00	0.00	0.00	0.02	0.10	0.07	0.08
6-7	0.00	0.00	0.00	0.00	0.00	0.00	0.00	0.00	0.01	0.02	0.02	0.02
7-8	0.00	0.00	0.00	0.00	0.00	0.00	0.00	0.00	0.00	0.00	0.00	0.00

Table A - 4 Wave climate schematization for scenario 2 of Sediment transport potential approach.

No. conditions	Dir. Range	Rep. dir. (^{0})	Rep. Hs (m)	Rep. Tp (s)	occ.%	Wind spd. (m/s)	Wind dir. (^{0})	Morfac
1	315-45	355	1.11	5.88	38.6	6.49	355	272.5
2	195-255	228	1.21	5.14	21.0	6.84	228	148.4
3	255-315	286	1.16	5.52	18.7	6.68	286	131.9
4	315-45	343	2.91	7.60	7.3	11.89	343	51.5
5	195-255	231	2.79	6.75	7.2	11.57	231	50.8
6	255-315	284	3.13	7.20	7.1	12.48	284	50.4

Table A - 5 Clusters of wave conditions selected for scenario 2 of Sediment transport potential approach (six dir. Bins and one wave height class)

%	N 345-15	NNE 15-45	ENE 45-75	E 75-105	ESE 105-135	SSE 135-165	S 165-195	SSW 195-225	WSW 225-255	W 255-285	WNW 285-315	NNW 315-345
0-1	7.19	4.87	2.26	1.02	0.74	0.83	1.17	3.33	4.81	3.69	4.38	6.33
1-2	6.09	2.78	1.91	0.75	0.43	0.35	0.81	4.79	5.80	4.06	4.52	7.12
2-3	1.34	0.51	0.38	0.07	0.01	0.03	0.11	1.89	2.89	2.04	1.93	2.88
3-4	0.24	0.07	0.01	0.00	0.01	0.00	0.02	0.39	0.99	0.78	0.79	0.97
4-5	0.04	0.00	0.00	0.00	0.00	0.00	0.00	0.03	0.18	0.31	0.29	0.31
5-6	0.01	0.00	0.00	0.00	0.00	0.00	0.00	0.00	0.02	0.10	0.07	0.08
6-7	0.00	0.00	0.00	0.00	0.00	0.00	0.00	0.00	0.01	0.02	0.02	0.02
7-8	0.00	0.00	0.00	0.00	0.00	0.00	0.00	0.00	0.00	0.00	0.00	0.00

Table A - 6 Wave climate schematization for scenario 1 of Sediment transport potential approach

No. conditions	Dir. Range	Rep. dir. (^{0})	Rep. Hs (m)	Rep. Tp (s)	occ.%	Wind spd. (m/s)	Wind dir. (^{0})	Morfac
1	-15-15	358	1.43	6.07	18.5	7.54	358	13.9
2	195-225	215	1.71	5.40	12.9	8.43	215	9.7
3	225-255	239	1.91	5.65	18.2	9.05	239	13.7
4	255-285	270	2.08	5.86	13.6	9.56	270	10.2
5	285-315	301	1.98	6.10	14.8	9.28	301	11.1
6	315-345	331	1.90	6.62	21.9	9.02	331	16.4

Table A - 7 Wave climate schematization for scenario 4 (summer) of Sediment transport potential approach

No. conditions	Dir. Range	Rep. dir. (^{0})	Rep. Hs (m)	Rep. Tp (s)	occ.%	Wind spd. (m/s)	Wind dir. (^{0})	Morfac
1	195-285	243	0.66	4.64	16.3	4.94	243	113.9
2	285-45	347	0.62	5.21	34.6	4.81	347	241.0
3	195-285	240	1.40	5.45	12.8	7.46	240	89.5
4	285-45	341	1.40	6.18	25.1	7.47	341	175.3
5	195-285	243	2.36	6.41	3.2	10.38	243	22.3
6	285-45	333	2.37	6.99	5.7	10.42	333	39.9
7	195-255	233	3.40	7.22	0.2	13.14	233	1.7
8	255-315	289	3.64	7.67	0.6	13.73	289	4.7
9	315-45	334	3.75	8.38	1.1	13.99	334	8.0

Table A - 8 Wave climate schematization for scenario 4 (winter) of Sediment transport potential approach

No. conditions	Dir. Range	Rep. dir. (^{0})	Rep. Hs (m)	Rep. Tp (s)	occ.%	Wind spd. (m/s)	Wind dir. (^{0})	Morfac
1	195-285	239	0.70	4.82	9.9	5.10	239	68.9
2	285-45	345	0.67	5.90	15.6	5.00	345	108.8
3	195-285	238	1.49	5.71	20.4	7.76	238	142.1
4	285-45	340	1.46	6.47	20.6	7.64	340	143.9
5	195-285	241	2.44	6.58	12.6	10.60	241	87.7
6	285-45	331	2.43	7.22	9.4	10.57	331	65.5
7	195-255	234	3.60	7.47	3.6	13.62	234	24.9
8	255-315	284	4.00	7.99	4.9	14.53	284	34.2
9	315-45	338	3.89	8.71	2.9	14.30	338	20.4

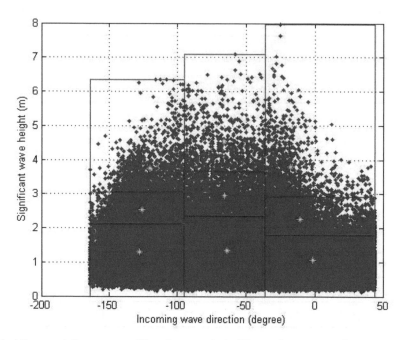

Figure A - 1 Representative wave conditions for scenario 1 of Energy flux approach (three directional bins and three wave height classes): boarder of the bins (red lines), all wave records (blue dots), representative wave conditions (green dots)

Table A - 9 Wave climate schematization for scenario 1 of Energy Flux approach

No. conditions	Dir. Range	Rep. dir. (^{0})	Rep. Hs (m)	Rep. Tp (s)	occ.%	Wind spd. (m/s)	Wind dir. (^{0})	Morfac
1	195-265	234	2.54	6.61	5.2	10.89	234	37.1
2	195-265	296	1.35	5.74	21.3	7.30	296	150.7
3	195-265	358	1.07	5.82	33.3	6.38	358	235.6
4	265-324	232	1.31	5.20	25.1	7.16	232	177.2
5	265-324	349	2.27	7.10	6.1	10.11	349	43.3
6	265-324	294	2.95	7.26	3.5	12.00	294	25.0
7	324-45	239	3.75	7.54	2.1	13.96	239	14.9
8	324-45	294	4.52	8.58	1.3	15.64	294	9.0
9	195-265	342	3.81	8.56	1.8	14.11	342.53	12.7

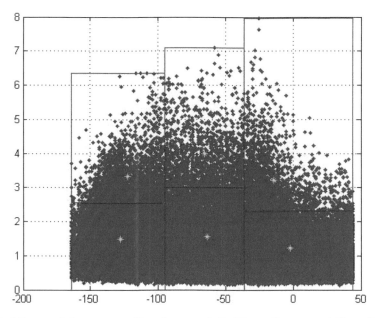

Figure A - 2 Representative wave conditions for scenario 2 of Energy flux approach (three directional bins and two wave height classes): boarder of the bins (red lines), all wave records (blue dots), representative wave conditions (green dots).

Table A - 10 Wave climate schematization for scenario 2 of Energy flux approach

No. conditions	Dir. Range	Rep. dir. (^{0})	Rep. Hs (m)	Rep. Tp (s)	occ.%	Wind spd. (m/s)	Wind dir. (^{0})	Morfac
1	195-265	232	1.49	5.3	28.3	7.74	232	199.5
2	195-265	237	3.33	7.2	4.2	12.98	237	29.7
3	265-324	296	1.56	5.9	23.6	7.97	296	166.3
4	265-324	294	3.99	8.1	2.6	14.53	294	18.4
5	324-45	357	1.23	5.9	37.3	6.91	357	262.9
6	324-45	345	3.23	7.9	4.0	12.72	345	28.7

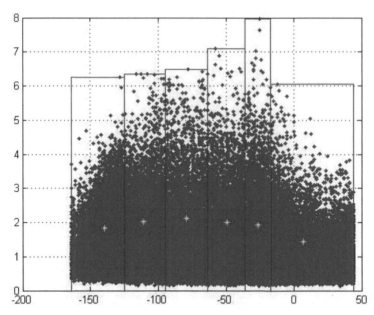

Figure A - 3 Representative wave conditions for scenario 3 of Energy flux approach (six directional bins and one wave height class): boarder of the bins (red lines), all wave records (blue dots), representative wave conditions (green dots)

Table A - 11 Wave climate schematization for scenario 3 of Energy flux approach

No. conditions	Dir. Range	Rep. dir. (^{0})	Rep. Hs (m)	Rep. Tp (s)	occ.%	Wind spd. (m/s)	Wind dir. (^{0})	Morfac
1	195-235	220	1.84	5.49	18.0	17.40	220	127.1
2	235-265	249	2.02	5.70	14.5	17.72	249	102.1
3	265-296	280	2.13	5.92	12.6	18.05	280	88.6
4	296-324	310	1.99	6.24	13.6	18.49	310	96.0
5	324-343	333	1.94	6.67	13.4	19.03	333	94.7
6	343-45	7	1.43	5.87	27.9	17.98	7	196.9

References

Blom A. 2008, Different approaches to handling vertical and stream-wise sorting in modeling river morphodynamics. Water Resource Res. 44:W03415. doi:10.1029/2006WR005474

Brady, A.J., Sutherland, J., 2001. COSMOS modelling of COAST3D Egmond main experiment. HR Wallingford Report TR 115.

Bruneau, N., Fortunato, A., Dodet, G., Freire, P., Oliveira, A., Bertin, X., 2011, Future evolution of a tidal inlet due to changes in wave climate, Sea level and lagoon morphology (Óbidos lagoon, Portugal), Continental Shelf Research, Volume 31, Issue 18, 15 November 2011, Pages 1915-1930, ISSN 0278-4343, 10.1016/j.csr.2011.09.001.

Bruun P., Mehta A.J., Johnsson I.G., 1978. Stability of Tidal Inlets. Theory and Engineering. Elsevier Scientific, Amsterdam, The Netherlands p. 506

Bruun, P., and Gerritsen, F. 1959. "Natural bypassing of sand at coastal inlets". Journal of the Waterways and Harbors, 85, 75-107.

C.E. Hanson, Eds., Cambridge University Press, Cambridge, UK, 315-356. NWO-ALW. 1999. Outer Delta Dynamics: process analysis of water motion, sediment transport and morphologic variability, Project proposal, Utrecht University, Utrecht.

Cayocca, F. (2001). Long-term morphological modeling of a tidal inlet: the Arcachon Basin, France. Coastal Engineering, 42, 115-142.

Davis, R. A., and Hayes, M. O. (1984). "What is a wave-dominated coast?" Marine Geology, 60, 313-329.

De Groot, T. (1999). Climate shifts and coastal changes in a geological perspective. A contribution to integrated coastal zone management, Geologie en mijnbouw, 77, 3651-361.

De Ronde, J. G., Van Marle, J. G. A., Roskam, A.P., and Andorka Gal, J.H., (1995). Golfrandvoorwaarden langs de Nederlandse kust op relatief diep water (in Dutch), Report RIKZ-95.024. Rijkswaterstaat RIKZ, The Hague.

De Swart, H.E. and Zimmerman, J.T.F. 2009. Morphodynamics of tidal inlet systems. Annual Review of Fluid Mechanics, 41 (2009), pp. 203–229

De Vriend, H. J., 1991. Mathematical modeling and large-scale coastal behavior, Part 1: physical processes. Journal of Hydraulic Research, 29, 727-740.

De Vriend, H. J., and Ribberink, J. S. 1996. Mathematical modeling of meso-tidal barrier island coasts. Part II: Process-based simulation models. Advances in Coastal and Ocean engineering, P. L.-F. LIU, ed., World Scientific Publishing Company., Singapore, 150-179.

De Vriend, H. J., Louters, T., Berben, F. M. L. and And Steijn, R. C., 1989. Hybrid prediction of intertidal flat evolution in an estuary. International Conference Hydraulic and Environmental Modeling of Coastal, Estuarine and Rivers waters, Bradford, U.K.

De Vriend, H.J., Capobianco, M., Chesher, T., De Swart, H.E., Latteux, B. and Stive, M.J.F., 1993. Approaches to long term modeling of coastal morphology: a review. Coastal Engineering 21 (1-3), 225-269.

Di Silvio, G., 1989. Modelling the morphological evolution of tidal lagoons and their equilibrium configurations. XXII Congress of the IAHR, Ottawa, Canada, 21-25 August 1989.

Dissanayake, D.M.P.K, 2011. Modelling Morphological Response of Large Tidal Basins to Sea Level Reise. UNESCO-IHE / TU Delft. Ph.D. Thesis, CRC Pres.

Dissanayake, D.M.P.K., Ranasinghe, R., Roelvink, J.A. 2012. The morphological response of large tidal inlet/basin systems to relative sea level rise Climatic Change, 113 (2), pp. 253-276.

Dissanayake, D.M.P.K., Roelvink, J.A., van der Wegen, M., 2009. Modelled channel patterns in a schematized tidal inlet. Coastal Engineering, 56 (11-12), pp. 1069-1083

Dobrochinski, J.P.H., 2009. Wave climate reduction and schematization for morphological modelling. MSc Thesis, Delft University of Technology, Delft, Univali University, Brazil.

Elias, E., Van der Spek, A., and Cleveringa, J., submitted, Morphodynamics of the Dutch Wadden Sea in the last century. Jur. Maine geology.

Elias, E. and Stive, M. The Effect of Stratification on the Residual Flow in a Mixed-Energy Tide-Dominated Inlet, Proceedings of coastal dynamics 2005, pp 1-13.

Elias, E. P. L., Stive, M. J. F., Bonekamp, J. G. and Cleveringa, J., 2003. Tidal inlet dynamics in response to human intervention. Coastal Engineering Journal, 45(4), 629-658.

Elias, E., 2006. Morphodynamics of Texel Inlet. IOS Press, The Netherlands

Elias, E.; Stive, M., and Roelvink, J.A., 2004. Impact of back-barrier changes on ebb-tidal delta evolution. Journal of Coastal Research, SI(42), 453-469. West Palm Beach (Florida).

Escoffier, F. F. 1940., The stability of tidal inlets. Shore and Beach, 114-115.

Eysink, W.D., 1990. Morphological response of tidal basins to change. Proc. 22nd Coastal Engineering Conference, ASCE, Delft, Juky 2-6, Vol. 2, the Dutch coast, paper no. 8, 1990, pp. 1948-1961.

Eysink, W.D., 1991. ISOS*2 Project : Impact of sea level rise on the morphology of the Wadden Sea in the scope of its ecological function, phase 1. Delft Hydraulic report H1300, Delft, The Netherlands.

Eysink, W.D.. 1992. ISOS*2 Project : Impact of sea level rise on the morphology of the Wadden Sea in the scope of its ecological function, phase 2, Delft Hydraulic report H1300, Delft, The Netherlands.

Fitzgerald, D. M. 1988. "Shoreline erosional-depositional processes associated with tidal inlets" Hydrodynamics and sediment dynamics of tidal inlets, Lecture Notes on Coastal and Estuarine Studies 29, Aubrey And D. G. Weishar, eds., pringer-Verlag, New York, 186-225.

Fitzgerald, D. M., Kraus, N. C., And Hands, E. B. 2000. Natural mechanisms of sediment bypassing at tidal inlets, Report ERDC/CHL CHETN-IV-30). US Army Corps of Engineers.

Fitzgerald, D. M., Penland, S., And Nummedal, D. 1984. "Control of barrier island shape by inlet sediment bypassing: East Frisian Islands, West Germany" Marine Geology, 60, 355-376.

Friedrichs, C. and Aubrey, D. G. ,1988. Non-linear tidal distortion in shallow well-mixed estuaries: a synthesis. Estuarine, Coastal and Shelf Sciences, 27, 521-545.

Galappatti, R. 1983. A depth integrated model for suspended transport, Communications on Hydraulics, vol. 83-7, Delft University of Technology, Delft, 114p.

Ganju, N. K., D. H. Schoellhamer, and B. E. Jaffe ,2009, Hindcasting of decadal-timescale estuarine bathymetric change with a tidal-timescale model, J. Geophys. Res., 114, F04019, doi:10.1029/2008JF001191

Gelfenbaum, G., Roelvink, J.A., Meijs, M., and Ruggiero, P., 2003. Process-based morphological modeling of Grays Harbor inlet at decadal timescales. Proceedings of the International

Conference on Coastal Sediments 2003. CD-ROM Published by World Scienti_c Publishing Corp. and East Meets West Productions, Corpus Christi, Texas, USA. ISBN 981-238-422-7.

Hands, E.B., and Shepsis, V., 1999. Cyclic channel movement at the entrance to Willapa Bay, Washington, U.S.A. Proceeding of Coastal Sediments '99, pp. 1522-1536.

Hayes, M. O. 1975. "Morphology of sand accumulation in estuaries.", Estuarine Research, L. E. CRONIN, ed., Academic Press, New York, 3-22.

Hayes, M. O., 1979. Barrier Island morphology as a function of tidal and wave regime, Barrier Islands: From the Gulf of St Lawerence to the Gulf of Mexico, S. P. Leatherman, ed., Academic Press, New York, 1-27.

Hibma, A. 2004. Morphodynamic modeling of channel-shoal systems, Communications on Hydraulic and Geotechnical Engineering, vol. 04-3, Delft University of Technology, Delft, 122p.

Hibma, A., De Vriend, H.J., Stive, M.J.F., 2003. Numerical modelling of shoal pattern formation in well-mixed elongated estuaries Estuarine, Coastal and Shelf Science, 57 (5-6), pp. 981-991. doi: 10.1016/S0272-7714(03)00004-0

Hirano M., 1971. River bed degradation with armouring. Trans Jpn Soc Civ Eng 3:194–195

Ikeda, S., 1982. Incipient Motion of Sand Particleson Side Slopes. Journal of the Hydraulics Division, ASCE 108 (1): 95–114.

Jarret, J.T., 1976. Tidal prism-inlet relationships, Gen. Invest. Tidal inlets Rep. 3, 32 pp, US Army Coastal Engineering and Research Centre. Fort Belvoir, Va.

Karssen, 1994a, 1994b

Karunarathna, H., Reeve, D., Spivack, M., 2008, Long-term morphodynamic evolution of estuaries: An inverse problem. Jur. Estuarine, Coastal and Shelf Science, Volume 77, 3, 385-395.

Karunarathna, H., Reeve, D., Spivack, M., 2009, Beach profile evolution as an inverse problem. Jur. Continental Shelf Research, Volume 29, 18, 2234-2239,

Kragtwijk N.G., Zitman T.G., Stive M., Wang Z.B. 2004, Morphological response of tidal basins to human interventions, Coastal Engineering, Volume 51, Issue 3, pp 207-221, doi: 10.1016/j.coastaleng.2003.12.008.

Krol, M., (1990), The method of averaging in partial differential equations, PhD thesis, University of Utrecht, 81 pp.

Latteux, B., 1995. Techniques for long-term morphological simulation under tidal action. Marine Geology, 126, pp 129-141.

Lesser, G. R., Roelvink, J. A., Van Kester, J. A. T. M. and Stelling, G. S., 2004. Development and validation of a three-dimensional model. Coastal Engineering, 51, 883-915.

Lesser, G.R. , 2009. An Approach to Medium-Term Coastal Morphological Modeling: UNESCO-IHE / TU Delft. Ph.D. Thesis, CRC Pres.

Lesser, G.R., De Vroeg, J.H., Roelvink, J.A., De Gerloni, M. and Ardone, V., 2003. Modeling the morphological impact of submerged offshore Breakwaters Proc.Coastal Sediments V03.

Louters, T., and Gerritsen, F. (1994). The Riddle of the Sands. A Tidal System's Answer to a Rising Sea Level., Report RIKZ-94.040. Rijkswaterstaat RIKZ, The Hague.

Marciano, R., Z.B. Wang, A. Hibma, H.J. de Vriend and Defina A., 2005, Modelling of channel patterns in short tidal basins. Journal of Geophysical Reserach, 100, F01001, doi:10.1029/2003JF000092.

Nahon A., Bertin X., Fortunato A.B., Oliveira A., 2012. Process-based 2DH morphodynamic modeling of tidal inlets: A comparison with empirical classifications and theories, Marine Geology, Volumes 291–294, pp 1-11, doi :10.1016/j.margeo.2011.10.001.

Nicholls, R.J., P.P. Wong, V.R. Burkett, J.O. Codignotto, J.E. Hay, R.F. McLean, S. Ragoonaden and C.D. Woodroffe, 2007: Coastal systems and low-lying areas. Climate Change 2007: Impacts, Adaptation and Vulnerability. Contribution of Working Group II to the Fourth Assessment Report of the Intergovernmental Panel on Climate Change, M.L. Parry, O.F. Canziani, J.P. Palutikof, P.J. van der Linden and

O'Brien, M.P., 1969. Equilibrium flow areas of inlets on sandy coasts, Journal of the Waterways and Harbours Division, ASCE. 95(WW1), pp. 43-51.

O'Brien, M. P. ,1931. Estuary tidal prisms related to entrance areas. Civil Engineering, 1, 738-739.

Oertel, G. F. (1975). "Ebb-tidal deltas of Georgia Estuaries", Estuarine research, L. E. CRONIN, ed., Academic press, New York, 267-276.

Oost, A. P. (1995). Dynamics and sedimentary development of the Dutch Wadden Sea with emphasis on the Frisian Inlet. A study of barrier islands, ebb-tidal deltas, inlets and drainage basins, Geologica Ultraiectina, Mededelingen van de Faculteit Aardwetenschappen no. 126, Utrecht University, 454 p.

Overeem, Van J., Stejin, R.C. And Van Banning, G.K.F.M., 1992. Simulation of morphodynamics of tidal inlet in the Wadden Sea. In: Sterr, H., Hofstede, J., and Plag, H., (Editors), Proceeding International Coastal Congress, Kiel, Peter Lang Verlag, Frankfurt am Main, pp. 351-364.

Ranasinghe, R., Swinkels, C., Luijendijk, A., Roelvink, D., Bosboom, J., Stive, M., Walstra, D., 2011, Morphodynamic upscaling with the MORFAC approach: Dependencies and sensitivities. Jur. Coastal Engineering, Volume 58, 8, 806-811.

Renger, E. and Partenscky, H. W., 1974. Stability criteria for tidal basins. Proc. 14th Coastal Engineering Conference, ASCE, Vol2, ch. 93,1974 pp. 1605-1618.

Reniers, A.J.H.M., Roelvink, J.A., Thornton, E.B., 2004. Morphodynamic modelling of an embayed beach under wave group forcing. Journal of Geophysical Research 109 (C01030).

Roelvink, D., Lesser, G. and Van der Wegen, M., Morphological modelling of the wet-dry interface at various timescales, proceedings ICHE conference 2006, Philadelphia, USA

Roelvink, J. A. and D. J. R. Walstra, 2004. Keeping it simple by using complex models. In Proceedings of the 6th International Conference on Hydro-Science and Engineering. Advances in Hydro-Science and Engineering, vol. VI, page p. 12. Brisbane, Australia. 217.

Roelvink, J.A. 2006. Coastal morphodynamic evolution techniques. Coastal Engineering 53, pp.277-287.

Roelvink, J.A. and Reniers A., 2012. A guid to mdeling coastal morphology. Advances in coastal and ocean engineering Vol 12. World Scientific Publishing. p. 274

Roelvink, J.A., Van der Kaaij R.M. Ruessink G. , 2001. Calibration and verification of large-scale 2D/3D flow models phase 1. Project No. Z3029.10 ONL Coast and Sea Studies.

Roelvink, J.A., Van Kessel, T., Alfageme, S. and Canizares, R. 2003. Modelling of barrier island response to storms Proc. Coastal Sediments V03.

Roelvink, J.A., Walstra, D.J.R., Chen, Z., 1994. Morphological modelling of Keta lagoon case. Proc. 24th Int. Conf. on Coastal Engineering. ASCE, Kobe, Japan.

Roskam, A. P. 1988. Golfklimaten voor de Nederlandse Kust (in Dutch), Report GWAO-88.046. Rijkwaterstaat RIKZ, The Hague.

Schwartz, M.L., 1973. Barrier Islands. Hutchinson & Ross, Dowden.

Sha L.P. 1989b. Variations in ebb-delta morphologies along the west and east Frisian islands, the Netherlands and Germany. Marine Geology, 89:11--28,

Sha L.P., 1989a. Cyclic morphologic changes of the ebb-tidal delta, Texel inlet, the Netherlands. Geology & Mijnbouw, 68:35--49,

Sha, L.P. and Van Der Berg, J.H. ,1993. "Variation in ebb-tidal delta geometry along the coast of Yhe Netherlands and German Bight". Journal of Coastal Research, 9(3), pp.730-746.

Schuttelaars, H. M. and De Swart, H. E., 1996. An idealized long-term model of a tidal embayment. European Journal Mechanics Fluids, 15, 55–80.

Soulsby D.H. 1997. Dynamics of marine sands. A manual for practical applications. Thomas Telford Ltd,

Speer, P.E., Aubrey D., 1985 A study of non-linear propagation in shallow inlet/estuarine systems. Part II, theory. Estuar Coast Shelf Sci 21:207–224.

Speer, P. E., Aubrey, D. and Friedrichs, C., 1991. Nonlinear hydrodynamics of shallow tidal inlet / bay systems. Tidal Hydrodynamics, B. B. Parker, ed., John Wiley, New York, 321-339.

Steijn, R., Roelvink, J.A., Rakhorst, D., Ribberink, J., and van Overeem, J., 1998. North Coast of Texel: a comparison between reality and prediction. Proc. 26th Int. Conf. On Coastal Engineering, Copenhagen, ASCE, New York, pp. 2281-2293.

Stive, M. J. F., and Eysink, W. D., 1989. Voorspelling ontwikkeling kustlijn 1990- 2090. fase3. Deelrapport 3.1: Dynamisch model van het Nederlandse Kustsysteem (in Dutch), Report H825. Waterloopkundig laboratorium, Delft.

Stive, M. J. F., M., C., Wang, Z. B., Ruol, P. and Buijsman, M. C., 1998. Morphodynamics of a tidal lagoon and adjacent coast. Proc. 8th International Biennial Conference on Physics of Estuaries and Coastal Seas, The Hague, 397-407.

Stive, M.J.F. and Z.B. Wang, 2003, Morphodynamic modelling of tidal basins and coastal inlets, In C. Lakkhan (Ed.) Advances in coastal modelling, Elsvier Sciences, pp. 367-392.

Steetzel, H., 1995 Voorspelling ontwikkeling kustlijn en buiten-delta's waddenkust over de periode 1990–2040. WL Report No. H1887 prepared for Rijkswaterstaat, The Hague, The Netherlands (in Dutch)

Sutherland, J., Hall, L.J. Chesher, T.J., 2001. Evaluation of the coastal area model PISCES at Teignmouth (UK). HR Wallingford Report TR125

Sutherland, J., Peet, A.H. and Soulsby, R.L., 2004. Evaluating the performance of morphological models. Coastal Engineering 51, pp. 917-939.

Sutherland, J., Soulsby, R.L., 2003. Use of model performance statistics in modelling coastal morphodynamics. Proceedings of the International Conference on Coastal Sediments 2003. CD-ROM Published by World Scientific Publishing. and East Meets West Productions, Corpus Christi, TX, USA, ISBN: 981-238-422-7.

T. Duong , R. Ranasinghe, A. Luijendijk, A. Dastgheib and D. Roelvink, 2012 climate change impacts on stability of small tidal inlets: A numerical modelling study using the Realistic Analogue approach. Proceedings of PIANC-COPEDC VIII , India, pp 594-602

Tung T.T., Stive M.J.F., Van de Graff J., Walstra D.J.R.2008. Morphological stability of tidal inlets using process-based modelling Proceedings of International Conference on Coastal Engineering 2008.

Tung T.T., Walstra D.J.R., Van de Graff J., Stive M.J.F., 2009. Morphological modelling of tidal inlet migration and closure. Journal of Coastal Research, SI56 , pp. 1080–1084

Tung, T.T., 2011. Morphodynamics of seasonally closed coastal inlets at the central coast of Vietnam, PhD Thesis, Faculty of Civil Engineering and Geosciences, Delft University of technology.

Van de Kreeke J., 2006. An aggregate model for the adaptation of the morphology and sand bypassing after basin reduction of the Frisian Inlet, Coastal Engineering, Volume 53, Issues 2–3, pp 255-263, doi: 10.1016/j.coastaleng.2005.10.013.

Van de Kreeke, J. and Robaczewska K., 1993. Tide induced residual transport of coarse sediment; application to the Ems estuary, Netherlands. Journal of Sea Research, 31 (3), Netherlands Institute for Sea Research, 209-220.

Van de Waal, R. 2007. Sediment transport pattern in the Dutch Wadden Sea. Delft Hydraulic report no. Z4169.00.

Van der Wegen M. 2010. Modeling morphodynamic evolution in alluvial estuaries. UNESCO-IHE / TU Delft. Ph.D. Thesis, CRC Pres.

Van der Wegen, M., Wang, Z.B., Savenije, H.H.G. and Roelvink J.A. 2008 Long-term morphodynamic evolution and energy dissipation in a coastal plain, tidal embayment, Journal Of Geophysical Research, Vol. 113, F03001, Doi:10.1029/2007jf000898, 2008

Van der Wegen, M. and Roelvink, J.A., 2008. Long-term estuarine morphodynamics evolution of a tidal embayment using a 2 dimensional process based model, Journal of Geophysical Research, 113, C03016, doi:10.1029/2006JC003983.

Van der Wegen, M., Dastgheib, A., Jaffe, B.E., Roelvink, D., 2011. Bed composition generation for morphodynamic modeling: Case study of San Pablo Bay in California, USA. Ocean Dynamics, 61 (2-3), pp. 173-186. doi: 10.1007/s10236-010-0314-2

Van Dongeren, A.D. and H.J. De Vriend (1994), A model of morphological behaviour of tidal basins, Coastal Engineering, 22, 287-310

Van Geer, P., 2007, Long - term morphological evolution of the Western Dutch Wadden Sea, Delft Hydraulic report Z4169, Delft, The Netherlands.

Van Rijn, L.C., Ruessink, B.G., Mulder, J.P.M. (Eds.), 2002. COAST3D—EGMOND, the Behaviour of Straight Sandy Coast on the Time Scale of Storms and Seasons. Aqua Publications, Amsterdam.

Van Rijn, L.C., Walstra, D.J.R., Grasmeijer, B., Sutherland, J., Pan, S., Sierra, J.P., 2003. The predictability of cross-shore bed evolution of sandy beaches at the time scale of storms and seasons using process-based profile models. Coastal Engineering 47, 295 – 327

Walton, T.L., Adams, W.D., 1976. Capacity of inlet outer bars to store sand, In: Proc. 15th Coastal Engineering Conf., Honolulu, ASCE, New York, Vol. II, pp. 1919-1937.

Wang Z.B., Hoekstra P., Burchard H., Ridderinkhof H., De Swart H., Stive M., 2012. Morphodynamics of the Wadden Sea and its barrier island system, Ocean & Coastal Management, Available online 5 January 2012,doi:10.1016/j.ocecoaman.2011.12.022.

Wang, Z. B., Louters, T. and De Vriend, H. J., 1995. Morphodynamic modeling for a tidal inlet in the Wadden Sea. Marine Geology, 126, 289-300.

Wijnberg, K. M. 1995. Morphologic behavior of a barred coast over a period of decades., Faculteit Ruimtelijke Wetenschappen, University Utrecht, Utrecht, 245 pp.

Exposure

Refereed Journals:

2011
Van der Wegen, M., Dastgheib A., Jaffe B., Roelvink D., "Generation of initial bed composition for morphodynamic hindcasting of hydraulic mining deposits in San Pablo Bay, California", Ocean Dynamics Journal, Volume 61, PP 173-186.

2010
Adigeria V., Dastgheib A., Boeriu P., "Modeling versus analytical calculation of the loading chambers" Nile water science and engineering journal, Volume 3, PP 90-96.

2010
Van der Wegen, M., Dastgheib A. and Roelvink J.A., "Morphodynamic modeling of tidal channel evolution in comparison to empirical PA relationship", Journal of Coastal Engineering, Volume 57, PP 827-837.

2008
Dastgheib A., Roelvink J.A., Wang Z.B. "Long-term process-based morphological modeling of the Marsdiep tidal basin" Journal of Marine Geology, Volume 256, PP 90-100

Proceedings and Presentations:

2012
A. Dastgheib and D. Roelvink. "Zuiderzee is now called IJsselmeer: Process-based Modeling". Proceedings of 20th NCK-days 2012, PP. 91-95.

2012
A. Dastgheib, M.R. Rajabalinejad, R. Ranasinghe and D. Roelvink. "A probabilistic approach to investigate the effect of wave chorology on process-based morphological modeling". Proceedings of PIANC-COPEDC VIII , India, pp 708-716

2012
T. Duong , R. Ranasinghe, A. Luijendijk, A. Dastgheib and D. Roelvink, "CLIMATE CHANGE IMPACTS ON THE STABILITY OF SMALL TIDAL INLETS: A numerical modeling study using the Realistic Analogue approach". Proceedings of PIANC-COPEDC VIII , India, pp 594-602

2010
Walstra D.J.R., J.A. Roelvink, J.E.A. Storms, N. Geleynse, T.T. Tung, M. Van der Wegen, A. Dastgheib, D.M.P.K. Dissanayake and R. Ranasinghe, Process-based long-term morphological modeling: the state of the art and the way ahead, Proc."Les littoraux à l'heure du changement climatique" in the journal "les Indes Savantes".

2010
M van der Wegen, D Roelvink, R Ranasinghe, A Dastgheib, P Dissanayake "Sea level rise and morphodynamic evolution of tidal basins". Deltas in Times of Climate Change 2010, Rotterdam, the Netherlands.

2010	D Roelvink, M van der Wegen, A Dastgheib, P Dissanayake & R Ranasinghe, "Predictability of the main channel and shoal pattern of estuaries and tidal inlets". Proceedings of the Fifteenth International Biennial Physics of Estuaries and Coastal Seas conference (PECS 2010), Sri Lanka PP. 190-197.
2010	Dastgheib A., Garae N., and Ligteringen H., "Port Development in Vanuatu" Port infrastructure seminar 2010, Delft, The Netherlands.
2010	Dastgheib A., Van der Wegen, M, Roelvink J.A., "Generating initial bed composition for long-term morphological modeling of tidal basins " 32nd International conference on coastal Engineering (ICCE 2010), Shanghai, China.
2010	Dastgheib A., Aburoweis M. Roelvink J.A. "Applying waves on the long-term morphological simulation of tidal basins" NCK Days 2010, Zeeland, The Netherlands
2009	Dastgheib A., Roelvink J.A., Van der Wegen M., "Sediment Mixtures in Long-Term Morphological Simulations of Tidal Basins" NCK Days 2009, Texel, The Netherlands
2009	Dastgheib A., Roelvink J.A., Van der Wegen M., "Effect Of Different Sediment Mixtures On The Long-Term Morphological Simulation Of Tidal Basins" Proceedings of RCEM 2009, Santa Fe, Argentina. PP 913-918
2008	Dastgheib A., Wang Z.B., de Ronde J. Roelvink J.A. "Modeling Of Mega-Scale Equilibrium Condition Of Tidal Basins In The Western Dutch Wadden Sea Using A Process-Based Model" Proceedings of PIANC-COPEDEC VII Conference 2008, Dubai
2008	Dastgheib A., Van der Wegen M., Dissanayake D.M.P.K., Roelvink J.A. "Long-Term Process-Based Morphological Modelling of Tidal Basins And Estuaries in The Netherlands" Proceedings of 31st International conference on coastal Engineering (ICCE 2008), Hamburg, Germany, PP. 2231-2242
2008	Van der Wegen M., Dastgheib A., Roelvink J.A. "Tidal inlet evolution along the Escoffier curve and empirical prism-cross section relationship using a 2D, process-based approach" 14th International Conference on Physics of Estuaries and Coastal Seas (PECS 2008) Liverpool, UK.
2008	Dastgheib A., Van der Wegen M., Roelvink J.A "Using a Process Based Model to Re-produce Escoffier Closure Curve" NCK Days 2008, Delft, The Netherlands
2008	Adigeria V.; Dastgheib A., Boeriu P., "Consideration about the design of the loading chamber of a small hydropower plant" HYDRO 2008 Conference, Ljubljana, Slovenia
2003	Dastgheib, A., Alaee, M.J. "Field Observation and numerical Modelling of FCSC – Rajaee Port , Iran" 6th International Conference on Coastal and Port Engineering in Developing Countries (COPEDEC VI), Colombo, Srilanka

2002	Alaee, M.J., Pattiaratchi C. , Dastgheib A. "Large Scour Holes Induced by Coastal Currents" (Poster) 11th International Conference on Physics of Estuaries and Coastal Seas (PECS 2002) Hamburg, Germany
2002	Dastgheib, A. , Alaee, M.J. "Numerical Modelling of Flow Structure in the Vicinity of Shahid Rajaee Port – Bandar Abbas – Iran" 5th International Conference on Coasts, Ports, and Marine Structures (ICOPMS 2002), Ramsar, Iran

Other:

2007	Dastgheib A. "Long-term Morphological Modeling of Marsdiep Basin in the Dutch Wadden Sea, the Netherlands" MSc. Thesis, UNESCO-IHE, WL I Delft Hydraulics
2001	Dastgheib, A. "A numerical study on the interaction between Flow-Curvature-Induced-Secondary-Circulation (FCSC) and sea bed" MSc. Thesis, Amir Kabir University of Tech, Tehran, Iran (in Persian)
2001	Dastgheib, A. "A Review on the Mechanisms of sea bed scouring" Amir Kabir University of Tech, Tehran, Iran

About the author

Ali Dastgheib was born in Shiraz, Iran, in 1978. He attended the NODET (National organization for development of exceptional talents) middle and high school from 1988 to 1995. He graduated with a B.Sc. in Civil engineering from Shiraz University, Iran in 1999. Later he received a Master of Science (Gold Medalist) in Civil Engineering - Hydraulic Structures from Amir Kabir University of Technology (Tehran Polytechnic), Iran in 2001. Then he worked for Port and Maritime Organization of Iran for one year as a researcher in coastal engineering department. In late 2002 he co-founded Pouya Tarh Pars Consulting engineers and worked as a design manager dealing with port master planning and design of coastal structures. In 2004 he became the head of coastal and port engineering department in Tarh Pars Consulting engineers. In April 2007 he graduated with another Master of Science in coastal engineering and port development from UNESCO-IHE, Delft after carrying out his graduation research in Deltares (known as WL | Delft Hydraulics at the time). Since October 2007 he is working at UNESCO-IHE as a lecturer in port planning and coastal engineering carrying out research and advisory projects in the field of coastal morphology, climate adaptation and coastal zone management. In UNESCO-IHE he has guided more than 15 students during their research period and given lectures on many diverse subjects including coastal engineering, morphological modeling, port planning and design, traffic simulation, game theory, etc.

T - #0419 - 101024 - C180 - 244/170/10 - PB - 9781138000223 - Gloss Lamination